BURLEIGH DODDS SCIENCE: INSTANT INSIGHTS

NUMBER 105

Carbon monitoring and management in forests

bd burleigh dodds
SCIENCE PUBLISHING

Published by Burleigh Dodds Science Publishing Limited
82 High Street, Sawston, Cambridge CB22 3HJ, UK
www.bdspublishing.com

Burleigh Dodds Science Publishing, 1518 Walnut Street, Suite 900, Philadelphia, PA 19102-3406, USA

First published 2024 by Burleigh Dodds Science Publishing Limited
© Burleigh Dodds Science Publishing, 2024, except the following: Chapter 3 remains the copyright of the author. All rights reserved.

Notice
No responsibility is assumed by the publisher for any injury and/or damage to persons or property as a matter of product liability, negligence or otherwise, or from any use or operation of any methods, products, instructions or ideas contained in the material herein.

British Library Cataloguing in Publication Data
A catalogue record for this book is available from the British Library

ISBN 978-1-83545-002-4 (Print)
ISBN 978-1-83545-003-1 (ePub)

DOI: 10.19103/9781835450031

Typeset by Deanta Global Publishing Services, Dublin, Ireland

Contents

Series list

Chapter 1

Optimizing forest management for soil carbon sequestration

Andreas Schindlbacher, Federal Research and Training Centre for Forests, Natural Hazards and Landscape (BFW), Austria; Mathias Mayer, Swiss Federal Institute for Forest, Snow and Landscape Research (WSL), Switzerland and University of Natural Resources and Life Sciences (BOKU), Austria; Robert Jandl, Federal Research and Training Centre for Forests, Natural Hazards and Landscape (BFW), Austria; and Stephan Zimmermann and Frank Hagedorn, Swiss Federal Institute for Forest, Snow and Landscape Research (WSL), Switzerland

1 Introduction

Forest carbon (C) cycling is a key component of the terrestrial C sink (Friedlingstein et al., 2020; Harris et al., 2021). C capture and storage in forest biomass and soils, therefore, has been identified as a keystone 'natural climate solution' (Bossio et al., 2020; Griscom et al., 2017) and forests have been assigned important roles in national and international CO_2 mitigation frameworks/strategies, such as the IPCC, the Bonn challenge, and the 4 per mill initiative (Minasny et al., 2017; Smith et al., 2014; Stanturf et al., 2019).

Forests store more than 90% of biomass C across terrestrial ecosystems and thus forest management can contribute to C sequestration (Bastin et al., 2019; Pan et al., 2011; Smith et al., 2016). However, practices have primarily been optimized to fulfill ecosystems services such as timber production, biodiversity, recreation, and natural hazard protection (e.g. Führer, 2000), while C storage in forest soils has received less attention, although it represents an important C pool, storing greater amounts of C than biomass in most eco-regions (Scharlemann et al., 2014).

http://dx.doi.org/10.19103/AS.2022.0106.18

Repeated soil inventories in Germany and France, for example, observed that forest soils have sequestered about 0.4 Mt C/ha/yr during the last few decades (Grüneberg et al., 2014; Jonard et al., 2017); this increase in C meets the targets of the 4 per mill initiative and increases the expectation that forest soils have high potential for C sequestration. On the other hand, climate change already affects forests, driving forest management towards adaptive measures, leading to the risk of sequestered C being re-released by accelerated soil organic matter (SOM) decomposition (Melillo et al., 2017; Prietzel et al., 2016; Schindlbacher et al., 2015) and/or by disturbances such as forest fires (Seidl et al., 2014).

The capacity for sustainable storage of soil organic SOC, therefore, is a moving target. Concepts for defining reference levels or benchmarks for stocks of organic C of specific forest soils have been proposed, but are difficult to validate or apply due to a lack of required soil characterization data (e.g. texture, clay content, aggregate size distribution, etc.) (De Vos et al., 2015; Rabot et al., 2018; Wiesmeier et al., 2019) and the high diversity of forest soil types in general (Fig. 1). Principally, SOC stocks are assumed to be in dynamic equilibrium under given environmental conditions (Böttcher and Springob, 2001; Schlesinger, 1990). This concept implies that if forest management

Figure 1 Forest soils are highly variable in soil types and organic carbon stocks. (a) Skeletic Stagnic Fluvisol, 73 tC ha^{-1} (18% in organic-layer); (b) Albic Podzol, 140 tC ha^{-1} (23% in organic-layer); (c) Folic Gleysol, 126 tC ha^{-1} (56% in organic-layer); (d) Ferralic Cambisol, 170 tC ha^{-1} (22% in organic-layer); (e) Calcaric Skeletic Cambisol, 140 tC ha^{-1} (5% in organic-layer). The displayed soils are examples from Central Europe, Austria. Photo credits: Rainer Reiter – BFW.

increases C inputs, SOC stocks increase until a new steady state is reached (Jandl et al., 2007). However, soils have only limited capacity to store SOC as the availability of reactive mineral surfaces for long-term C stabilization is limited and once sorption sites are occupied with organic matter, only a small fraction of C input residues may remain in the soil for longer time periods (Kaiser and Guggenberger, 2003). Consequently, C turnover increases with increasing SOC concentrations and C inputs are balanced out by SOC outputs, which leads to so-called C saturation (Stewart et al., 2007), with the maximal SOC stocks depending on site conditions (texture, mineralogy, and climate). In comparison to agricultural soils, forest soils with inherently high SOC stocks appear close to C saturation, but quantitative evidence is limited.

Forest management practices are primarily dedicated to optimizing wood and biomass production, but these practices also impact soils by altering the quantity and quality of C inputs, changing microclimatic conditions by modifying the light and water regime, and by physically disturbing soils during management operations (Jandl et al., 2007; Mayer et al., 2020). In the following sections, we present information on the effects of forest management on soil C sequestration and discuss whether management of forests (e.g. choice of tree species, harvest, density regulation and thinning, fertilization, etc.) can be optimized under expected climate change conditions. We also discuss afforestation and reforestation options and highlight the role of disturbance-management. Furthermore, we present results from a case study in Central Europe investigating forest soil C storage in mountain regions of Switzerland and Austria.

2 Forest management and soil carbon sequestration

2.1 Afforestation and reforestation

Afforestation of formerly non-forested land and reforestation of deforested land have been highlighted as important strategies to reduce atmospheric CO_2 (Bastin et al., 2019; Griscom et al., 2017). Although the majority (>90%) of the CO_2 is thought to be sequestered in the new forest biomass, a significant CO_2 sink has been attributed to C sequestration in soils (Bossio et al., 2020; Nave et al., 2018). Whether this soil C sequestration potential is realistic remains uncertain since most case studies demonstrate that both afforestation and reforestation do not necessarily result in increased SOC stocks (Mayer et al., 2020).

Cropland SOC pools are typically depleted by persistent agricultural biomass removal. After re- or afforestation, organic matter accumulates until the soil reaches a new equilibrium between C input (litter fall, rhizodeposition) and C output (respiration, leaching). Accordingly, afforested cropland soils have been shown to typically sequester C across various climatic zones (Ashwood et al., 2019; Don et al., 2011; Poeplau and Don, 2013). Soil C sequestration is generally faster in warmer (topical and sub-tropical) climatic regions (Don et al., 2011; Silver et al., 2000), but reforested croplands in temperate and hemi-boreal regions can also recover SOC stocks close to those of managed or natural forest soils within 100 years or less (Ashwood et al., 2019). Cropland afforestation in temperate and cold biomes first leads to an accumulation of litter and forest floor, particularly if conifers are planted. After a period of years to decades, the sequestered C enters the mineral soil C pool (Bárcena et al., 2014a,b). Faster increases in mineral soil C were observed in reforested tropical cropland (Silver et al., 2000). As croplands are a limited land resource, global population growth and food security do, however, constrain their use for reforestation (Griscom et al., 2017).

Grasslands have been proposed as favorable afforestation targets due to their large spatial distribution. In tropical and sub-tropical regions, reforested C-depleted pastures proved to sequester C in soil, but to a lesser extent as reforested croplands (Don et al., 2011; Silver et al., 2000). In temperate and colder biomes, the situation is different. Temperate grassland soils typically store similar or even higher amounts of C than temperate forest soils. This already indicates that the capacity to sequester additional C following reforestation is limited. Reforestation of European pastures had only modest effects on SOC and often led to, at least transient, soil C losses (Hiltbrunner et al., 2013; Poeplau and Don, 2013). A further common observation is that reforestation of temperate grasslands reshuffles the C density within the soil profile. Carbon content often decreases in deeper mineral soil layers and increases in the organic layers and forest floor (Hiltbrunner et al., 2013; Mobley et al., 2015). The SOC that accumulates at the soil surface is more labile and less protected than mineral-bound SOC in deeper soil layers (Poeplau and Don, 2013) and therefore could decompose faster if environmental conditions become more favorable for microbial activity (e.g. under climate warming) or if soil is disturbed (e.g. after harvest). Afforestation of native grasslands can cause considerable habitat and biodiversity loss (Temperton et al., 2019; Veldman et al., 2019) and has recently been shown to decrease soil carbon (Santos et al., 2020).

In Northern Europe, many upland sites with thick organic soil layers have been afforested with conifers. Lowering the water table through drainage and site preparation prior to tree planting typically leads to substantial C losses from the aerated peat layer (Simola et al., 2012). However, it has been suggested that initial C losses following afforestation can be compensated by C input throughout several tree rotations, and that afforestation will sequester C in afforested organic soil in the long term (Vanguelova et al., 2019). Recent studies again highlight the risks associated with tree planting on organic soils.

Planting of native tree species on organic moorland at four sites in Scotland, for example, resulted in soil C losses on a decadal timescale (Friggens et al., 2020). Similarly, afforested peatland soils at eight different locations across Ireland have been shown to act as net C sources to the atmosphere (Jovani-Sancho et al., 2021; Friggens et al., 2020), suggesting that, even if there is no change in peatland water tables, tree roots shift the fungal/mycorrhizal community composition in the subsoil and thereby could trigger an increased decomposition of organic matter.

An extensive study in Chinese forests, comparing more than 600 control and afforested plot pairs, clearly confirmed that afforestation effects depend on SOC stocks of the former land (Hong et al., 2020). At low SOC stocks, planting forests leads to C gains in the soil, while at high SOC stocks (>100 t SOC/ha), such as in grasslands and organic soils, afforestation caused SOC losses. Overall, C sequestration in soils was much lower than expected when extrapolated to the total investigation area (Northern China, 120 000 km^2), suggesting that the global soil C sequestration potential of re- and afforestation may be smaller than in the above-mentioned optimistic scenarios. It further highlights the importance for forest managers to consider soil properties/processes when implementing re- or afforestation measures. Carbon gains by newly forming forest biomass can be partially or even totally offset if trees are not planted at appropriate sites. It should also be considered that tree plantations experience higher water use than non-forest vegetation. This can significantly affect site-, catchment-, and landscape-hydrology and potentially increase soil salinization and acidification (Jackson et al., 2005).

2.2 Management of tree species

Planting tree species is an active measure to enhance forest productivity, produce specific wood, promote biodiversity and enhance soil fertility. Tree species have distinct impacts on soil C cycling by altering the quantity and quality of above- and below-ground inputs and by affecting below-ground communities associated with tree species (Peng et al., 2020; Reich et al., 2005; Vesterdal et al., 2013). These processes are frequently interwoven and although tree species might have strong effects on individual soil processes, observed net effects on SOC stocks are often relatively small (Grüneberg et al., 2019; Wiesmeier et al., 2013).

In general, broadleaf trees have deeper rooting systems and litter that is more easily decomposed, thereby favoring biological activity. These species provide greater C inputs into mineral soils as compared to most conifer trees, which show litter accumulation at the soil surface. Below-ground C inputs decompose more slowly than above-ground litter (Hansson et al., 2010), and as they interact also more strongly with mineral surfaces, root-derived C is assumed to be of primary importance for the formation of stable SOM (Rasse et al., 2005; Schmidt et al., 2011). Tree species may also influence SOC storage by their association with distinct mycorrhizal fungi. SOC stocks and vertical SOC distribution were found to be greater under trees associated with arbuscular mycorrhiza than under those associated with ectomycorrhiza (ECM) (Craig et al., 2018; Peng et al., 2020). The negative effect of ectomycorrhizal fungi on SOC stocks is attributed to the secretion of powerful oxidizing enzymes which allow the acquisition of nutrients bound to SOM but promote SOM decomposition (Zak et al., 2019).

Due to the greater rooting depth, increasing the uptake of base cations from subsoils, and to the annually renewed foliage, litter from broadleaf trees generally has a higher quality with a greater nutrient content than litter from conifer trees (e.g. Berger et al., 2006; Reich et al., 2005). Litter quality is associated with a distinct microbial community composition and functioning that may in turn affect SOC storage (Heděnec et al., 2020). Moreover, earthworms profit from calcium-rich litter of broadleaf species (Reich et al., 2005; De Wandeler et al., 2018). As earthworms transfer litter C into the mineral soil, broadleaf trees species are associated with smaller C stocks in the forest floor (Desie et al., 2020), but greater C stocks in the mineral soil and, thus, potentially sustainable C sequestration (Jandl et al., 2007; Mayer et al., 2020).

The effects of tree species frequently overlap with other site conditions (Grüneberg et al., 2019). For instance, spruce naturally grows in a colder and more humid climate with inherently greater SOC stocks than broadleaf trees, which impedes the direct identification of tree species effects. However, common garden experiments and soil C inventories observed pronounced tree species effects in forest floors with greater C stocks under coniferous trees due to reduced litter decomposition as compared to high-quality litter from broadleaf trees (Reich et al., 2005; Vesterdal et al., 2013; Wiesmeier et al., 2013). The opposite pattern exists for mineral soils with greater SOC stocks under broadleaf forests often associated with arbuscular mycorrhiza (Peng et al., 2020; Wiesmeier et al., 2013). However, although stabilization of SOC is generally enhanced by the interaction with reactive mineral surfaces, indicating that soils from broadleaf forests have greater potential to sequester C in the long-term, observed differences in total SOC stocks (organic layer + mineral soil) are rather small (Grüneberg et al., 2019; Wiesmeier et al., 2013). One reason for these relatively small tree species effects could be 'priming' by rhizodeposition where fresh C inputs in the mineral soils stimulate microbial activity, which in turn accelerates decomposition of old SOC (Dijkstra et al., 2021).

Management options to promote tree species have to consider the adaptation of tree species composition to ongoing climatic changes. For instance, as spruce responds sensitively to drought and becomes vulnerable to bark beetle infestations, promotion of spruce in low elevation forests in Central Europe will be inappropriate in a future climate with more frequent and severe drought (Hlásny et al., 2021; Jandl, 2020). Silver fir and oak, by contrast, form deeper rooting systems (Vitasse et al., 2019), which potentially provide a greater C input into the mineral soil (Reich et al., 2005). Planting Douglas fir, another drought-resistant conifer originating from North America, may conflict with the paradigm of basing forests on native tree species.

2.3 Harvest and harvest residues

The harvesting of timber is one of the most important management activities in forests worldwide. Harvesting or logging operations commonly focus on the stem section of a tree, as it yields the highest merchantable value. Harvest residues such as leaves/needles, twigs, branches, or stumps, have a comparable low economic value and are often left on-site following harvesting operations. However, growing demand for biofuels as a renewable energy source has also led to an interest in utilizing harvest residues (Popp et al., 2014). Harvesting and the utilization of harvest residues strongly affects the quantity and quality of organic matter input to soil. Moreover, decomposition processes may be altered by more favorable microclimatic conditions (e.g. warmer and moister), shifts in microbial community composition, and/or soil disturbance (e.g. compaction, erosion). These changes may result in a misbalance of C input and C output, which ultimately impacts forest soil C stocks (Jandl et al., 2007; Mayer et al., 2020). To maintain the long-term C sequestration potential of forest soils (e.g. over several rotation periods) it is therefore crucial to select harvesting practices that minimize net soil C losses.

Global meta-analyzes investigating the effects of forest harvesting operations on soil C storage (mainly by clearcut logging of the entire tree layer) revealed diverging results; soil C stocks following harvest have been shown to decrease when compared to unharvested control stands (Achat et al., 2015; James and Harrison, 2016; Nave et al., 2010) or to be unaffected (Hume et al., 2018; Johnson and Curtis, 2001). Comprehensive meta-analysis (James and Harrison, 2016) (112 studies), however, reports harvest-related C losses of −30%, −3%, and −18% in organic, mineral topsoil (0-15 cm depth), and deep soil horizons (60-100+ cm), respectively. Greater negative harvesting effects on C stocks in organic horizons than in mineral soil horizons were also shown (Achat et al., 2015; Nave et al., 2010). Carbon losses from organic horizons were generally larger in hardwood forests (−36%) than in coniferous or mixed forests (−20%) (Nave et al., 2010). Following a period of net C losses, soil C stocks in organic and mineral soil horizons begin to recover after 10 to 50 years following forest harvest (Achat et al., 2015; James and Harrison, 2016; Nave et al., 2010; Sun et al., 2004; Tang et al., 2009).

The impact of harvesting operations on soil C stocks has been shown to depend on the amounts of harvest residues left on site. Meta-analyzes showed soil C stocks were unchanged following stem-only (i.e. bole fraction) harvest (Hume et al., 2018; James et al., 2021) and increased (+18%) when compared to control stands (Johnson and Curtis, 2001). By contrast, whole-tree harvest (including branches, leaves/needles) decreased soil C stocks by −15% (James et al., 2021). Achat et al. (2015) showed larger C losses in the organic horizon (10–45%) and in deeper mineral soil horizons (10%) after harvest when residues were removed. This meta-analysis also found a larger effect of harvest residue removal on soil C stocks in temperate than in boreal forests (Achat et al., 2015). Other studies, however, did not find clear evidence that whole-tree harvest would decrease soil C stocks (Hume et al., 2018; James and Harrison, 2016). In a Sitka spruce forest in the UK, C stocks of apeaty gley soil were even higher after whole-tree harvest when compared to stem-only harvest (Vanguelova et al., 2010). This finding was related to higher decomposition rates underneath residues. Stump harvesting has been shown to strongly decrease C stocks in organic horizons across a range of forests in Scandinavia, while mineral top- and sub-soil horizons were unaffected (Clarke et al., 2021). In a temperate forest in Washington, USA, stump harvesting resulted in 24% lower soil C stocks (Zabowski et al., 2008). However, others found either no reduction in soil C stocks (Jurevics et al., 2016; Strömgren et al., 2013) or only a small decrease following stump harvest (Hyvönen et al., 2016; Kaarakka et al., 2016).

Whether partial harvesting practices that retain living trees on site would reduce logging-related soil C losses remains inconclusive. Shelterwood harvest in a hardwood forest in New England, USA, for example, resulted in a decrease in mineral soil C stocks, while C stocks in the litter layer were unaffected (Warren and Ashton, 2014). In a Chilean Lenga forest, soil C stocks decreased for 3 years after shelterwood harvest, but recovered 8 years after logging (Klein et al., 2008). Other studies report little or no differences in soil C stocks when comparing partial to conventional clearcutting practices (Christophel et al., 2015; Hoover, 2011; Puhlick et al., 2016).

The effects of harvesting operations on soil C stocks have been suggested to be ameliorated if harvest rotation periods are extended (Law and Waring, 2015; Noormets and Nouvellon, 2015) or soil disturbance from, for example, machinery movements is reduced (Achat et al., 2015). Machinery-related soil disturbance in particular is supposed to strongly enhance harvesting effects on soil C stocks (James and Harrison, 2016; Mayer et al., 2021). These negative effects may be reduced by selecting harvesting systems that alleviate soil disturbance (e.g. harvesting during winter when soils are frozen or snow-covered, cable-yard or helicopter logging) (Mayer et al., 2021).

2.4 Density regulation and thinning

Density regulation already starts with tree planting if natural tree regeneration is lacking or undesirable. Figure 2 shows the results of a density regulation trial in an Austrian spruce forest in which saplings were planted in various densities on an arable field. No differences were found in total soil C stocks among the different density treatments 30 years after planting (Fig. 2). Na et al. (2021), however, found a significant increase in total SOC stocks with increasing planting densities in 8-year-old larch plantations, concurrent with significantly increasing fine root biomass and above-ground litter input.

Once a forest stand is growing, thinning (the removal of designated trees to increase space for the remainder) is the traditional silvicultural practice used to maximize merchantable timber volume and financial value (Gonçalves, 2020).

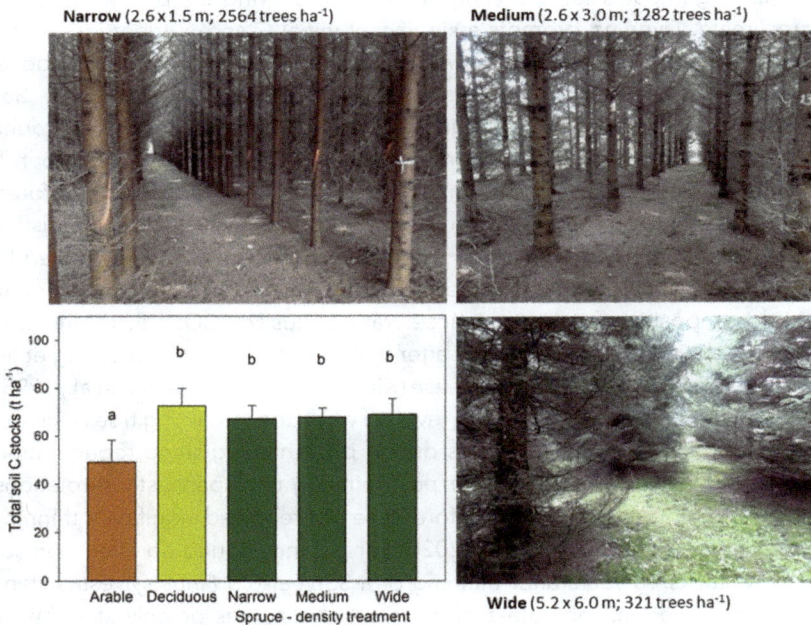

Figure 2 Soil carbon stocks in a density regulation trial by the Austrian Forest Research Centre – Institute of Forest Growth. Norway spruce (*Picea abies*) was planted in 1992 in narrow, medium and wide spacing. Deciduous trees (*Acer pseudoplatanus* and *Carpinus betulus*) were planted with a single spacing of 2.6 × 1.5 m. All trees were planted on an arable field. Total soil C stocks (including forest floor and litter) down to 40 cm soil depth were assessed in autumn 2020. Different letters above the individual bars indicate statistically significant differences (ANOVA, $p < 0.05$, n = 3–4 plots; 50 × 50 m each). While the soil C stocks of all forest plots were significantly higher than those of the adjacent arable field, there was no difference between deciduous and spruce stands, nor between the different density treatments.

Thinning is applied at various intensities, standages, intervals, and by different approaches (from above, below, selective, etc.). Thinning is also a management measure to reduce vulnerability to climate changes such as drought (Sohn et al., 2016), fire (Finkral and Evans, 2008; Fulé et al., 2001), and to facilitate structural and species diversity (D'Amato et al., 2013). Thinning always leads to a temporary reduction in stand density and therefore to a temporary reduction in above-ground biomass C stocks. After thinning, increased growth of the remaining trees can partly or fully compensate for the thinning biomass C loss – depending on thinning intensity and time elapsed thereafter. Thus above-ground biomass C stocks of thinned stands were often found to be lower than those of unmanaged stands (Bravo-Oviedo et al., 2015; Nilsen and Strand, 2008; Powers et al., 2011; Ruiz-Peinado et al., 2016) but they were also observed to fully regain above-ground C stocks of unmanaged stands in the longer run (Powers et al., 2012).

Thinning impacts soil C cycling primarily by altering C inputs and the soil microclimate. Thinning prompts a significant initial C and nutrient input from onsite harvest residues (except for whole trees harvest) and dead roots. Above-ground litterfall typically decreases after thinning (Zhang et al., 2018). Soil temperatures increase due to the increased solar penetration of the ground surface (Zhang et al., 2018). The magnitude of the temperature increase depends on the thinning intensity and to what extent the remaining understory or forest floor vegetation contributes to ground shading (Li et al., 2021). Soil moisture is mostly unaffected by thinning since increased transpiration is balanced by overall reduced water uptake. Under an improved soil microclimate, litter and SOM decomposition can become accelerated. Thus soil CO_2 effluxes increase in general during the first few years after thinning (summarized in Zhang et al., 2018) and forest floor C stocks decrease (Kim et al., 2018; Vesterdal et al., 1995).

Initial thinning effects, however, level off when the remaining trees increase their growth and the canopy closes during post-thinning stand regeneration. Most studies encompassing longer post-thinning timespans, stand-rotations, or even whole tree lifespans, therefore have not reported significant thinning effects on soil C stocks. Lim et al. (2020), for instance, found no effects on soil C and N stocks 18 years after thinning of a 35-year-old *Pinus sylvestris* stand in northern Sweden, regardless of whether whole trees or only stems were harvested (sampling depth 20 cm). Nilsen and Strand (2008) did not detect any significant effects on humus and mineral soil (sampling depth 100 cm) C storage 33 years after thinning of a Norway spruce stand in southern Norway either. No significant effects of various intensity repeated-thinning treatments over 50+ years were detected in the forest floor and mineral soil (sampling depth 30 cm) of a red Pine stand in Minnesota, and a northern hardwood forest in Wisconsin, USA (Jurgensen et al., 2012; Powers et al., 2011). However, in a pine stand in the Great Lakes region, Canada, almost 90 years of repeated intensive thinning not only substantially reduced above-ground biomass C but also reduced soil C stocks (sampling depth 30 cm) as well (Dávila Reátegui et al., 2021). In a global meta-analysis, Zhang et al. (2018) did not observe an overall significant effect of thinning on soil C stocks, though heavy thinning tended to decrease total soil C, whereas light thinning increased total soil C stocks. A similar trend toward soil C losses under heavy thinning and soil C gain under light and moderate thinning was observed by Gong et al. (2021) when meta-analyzing thinning effects in Chinese forest plantations.

Overall, the rather indifferent effects on forest soil C stocks do not indicate a certain way of adapting specific thinning strategies to optimize soil C sequestration. The role of thinning in future forest soil C sequestration should not, however, be underestimated. Thinning typically increases the resistance of the remaining trees to disturbances such as windthrow, snow pressure and drought, and allows for adaptive adjustments in tree species composition and vertical stand structure. Therefore, thinning can be a proper silvicultural measure to protect forest (soils) from disturbance-related C loss (see below). That such disturbances are mostly absent in the relatively rare published longer-term thinning trials might be an indication of a publishing bias toward experiments conducted at 'safe sites'. Accordingly, the role of thinning on the landscape scale might be somewhat underrated. As an example, Gong et al. (2021) reported an overall positive effect of thinning on the soil C stocks of more than 200 Chinese forest plantations – with the most pronounced effects in drought-prone areas.

2.5 Fertilization, liming and wood-ash application

Fertilization represents a potential measure to overcome nutrient deficiency, thereby promoting forest productivity. Fertilization impacts have most intensively been studied for N, and mechanisms include increased litter input through higher forest productivity and reduced SOM decomposition (Mayer et al., 2020). The latter can be attributed (1) to N-induced shifts in the decomposer community and reduced extracellular enzyme activities, (2) to a preferential substrate use shifting from recalcitrant SOM with high N content to fresh substrate with low N content, and/or (3) by adding N required for forming microbial necromass (e.g. Fog, 1988; Hagedorn et al., 2003). The review by Nave et al. (2009) indicated that N fertilization increases total soil C stocks (combined forest floor and mineral soil) by 7.7%, but the magnitude likely depends on the inherent N status of the soil. These benefits must be weighed against the associated environmental costs, such as nitrate leaching, soil acidification, enhanced soil N_2O emissions, and loss of biodiversity. Moreover, the production, transport, and application of synthetic fertilizers all entail fossil fuel combustion and emission of CO_2.

Liming aims at reversing soil acidification and improving soil fertility (Hildebrand, 1996). In principle, liming influences SOC storage by potentially increasing C inputs into soils through improved tree growth and a deeper rooting system (Schäffer et al., 2001). However, it also accelerates SOM decomposition (particularly in the forest floor) by enhancing C solubility and stimulating biological activity associated with an increased soil pH and a higher litter quality with higher Ca-contents (Kreutzer, 1995). Liming enhances the amount and activity of earthworms (Schäffer et al., 2001), which leads to a translocation of litter-derived C into the mineral soil and hence to a C loss in the forest floor but a potential C gain in the mineral soil. This C transfer is promoted by enhanced leaching of dissolved organic C through the rising pH (Feger et al., 2000).

Soil surveys and liming experiments show that liming effectively transforms thick forest floors into more 'active' humus forms (e.g. Court et al., 2018), which leads to SOC losses in the forest floor (e.g. Bauhus et al., 2004; Table 1). However, observed effects on SOC stocks in the mineral soil under the forest floor are inconsistent and it remains uncertain whether the C losses in the forest floor are outbalanced by C gains in the mineral soil (Table 1). The small effect sizes in the mineral soil could be related to stimulated SOC mineralization by liming, the C stocks in the mineral soil being already close to saturation, or the difficulty in detecting changes in the large C reservoir of mineral soils. While liming experiments tended to result in negative effects on total SOC stocks, SOC changes between repeated surveys of 385 limed sites in German forests showed slightly positive (+0.18 tC/ha/yr) liming effects (Grüneberg et al., 2019). The authors explained their findings by potentially deeper rooting trees and stabilization of SOM by Ca^{2+} added with the lime. Potentially, the enhanced C transfer into the mineral soil leads to a long-term stabilization of soil C.

The magnitude of liming responses appears to depend on site and soil conditions with greater SOC losses to be expected in soils with a thick forest floor. Potential C gains in soil C storage have to be outweighed against the release of CO_2 from carbonate in the lime (~12% of its mass), which corresponds to a release of 0.036 t CO_2–C/ha/yr at a standard addition of 3 t lime/ha every 10 years under temperate climate conditions. Moreover, liming alters microbial community composition (e.g. of ectomycorrhizal fungi (Kjøller and Clemmensen, 2008; van der Linde et al., 2018)) and bears the risk of increased NO_3^- leaching due to stimulated SOM and N mineralization in combination with enhanced nitrification (transformation of NH_4^+ to NO_3^-) at higher pH values. Potential negative effects can be minimized by the application of lime on small areas only, following shelterwood cutting interventions and by application of lime with low solubility.

Wood-ash application aims at returning base cations removed with harvesting back to the soil. Mechanistically, impacts of wood ash on soil

Table 1 Effect of liming on SOC storage in European forest soils

		Mode of liming	Years after liming y	Forest floor tC/ha	Forest floor %change	Mineral soil (0–20 cm) tC/ha	Mineral soil (0–20 cm) %change	Total soil tC/ha/y	Comments and limitations
Experiments									
Bauhus et al. (2004)	Beech (Solling) Stand	Dolomite (3t/ha)	8	-0.5	-3%	-14.4	-22%	-1.86	
	Beech (Solling) Gap	Dolomite (3t/ha)	8	-9.0	-60%	20.6	38%	1.45	
Kreutzer (1995)	Spruce (Höglwald)	Dolomite (4t/ha)	7	-7.2	-23%	0.8	4%	-0.91	0–5 cm depth; no data on deeper soil
Court et al. (2018)	Beech (N-France)	CaCO3 (2.5 t/ha)	20–40	-2.1	-0.1	n. sign.	n. sign.		Average of five sites
Marschner and Wilczynski (1991)	Spruce (Berlin)		3	-6.9	-23%	-2.9	-5%	-3.27	Heterogenous sandy site, liming was combined with K fertilization
Matzner et al. (1985)	Spruce (Solling)	CaCO3 (5 t/ha)	10	-7.1	-15%	-12.5	-19%	-1.96	Combined with N-fertilizer
	Beech (Solling)	CaCO3 (5 t/ha)	10	4.4	18%	20.6	37%	2.50	Combined with N-fertilizer
Persson et al. (1995)	Spruce (Sweden)	CaCO3 (9 t/ha)	42	-6.0	-40%	3.0	4%	-0.07	Includes a stand rotation
(values 0–50 cm)	Spruce (Sweden)	CaCO3 (12 t/ha)	37	-10.3	-30%	-10.7	-12%	-0.57	Planted former heathland
	Spruce (Sweden)	CaCO3 (10 t/ha)	38	-7.5	-94%	-8.0	-8%	-0.41	Includes a stand rotation
	Beech (Sweden)	CaCO3 (10 t/ha)	38	-20.0	-80%	0.0	0%	-0.53	
Soil survey									
Grüneberg et al. (2019)	German forests	mostly 3 t/ha	variable	-0.25 tC/ha/y		0.43 tC/ha/y		0.18 tC/ha/y	Large data set Comparability of sites uncertain

processes are similar to those of lime, with an increased soil pH and an improved supply with base cations (review by Huotari et al., 2015). lthough 'net effects on soil C balance have not been thoroughly evaluated' (see Huotari et al., 2015), in agreement with liming studies, C and N mineralization rates were found to be stimulated by wood-ash addition (Rosenberg et al., 2010). In a field study in Switzerland, Zimmermann and Frey (2002) observed long-lasting increases in soil CO_2 effluxes and a decrease in SOC content in the uppermost 5 cm by approximately 20% during the first 3 years after wood-ash application. Whether the C losses in the forest floor and upper mineral soils are compensated for by C gains in the deeper mineral soil through a C transfer mediated by soil fauna or by the development of a deeper rooting system remains unstudied.

2.6 Disturbance control

Forests around the world are increasingly being damaged by natural disturbance agents such as wildfires, droughts, pest outbreaks, or windthrows (Allen et al., 2010; Seidl et al., 2017; Senf et al., 2018; Williams et al., 2016). Climate change is considered to reinforce disturbance frequency and severity, and model predictions suggest that natural forest disturbance will be even more pronounced in the next few decades (Seidl et al., 2014). Natural disturbance can strongly impact soil C stocks of forest stands. Global meta-analyzes indicated that soil C concentrations and storage generally decrease following disturbance by fire, insects, and wind (Nave et al., 2011; Thom and Seidl, 2016; Wang et al., 2012; Zhang et al., 2015). Severe wildfires can cause a rapid loss in forest floor C by direct combustion of organic matter (Bowd et al., 2019; Nave et al., 2011; Pérez-Izquierdo et al., 2021). Forest ecosystems with thick organic layers (e.g. boreal forests) may thus be more vulnerable to C losses after fire than ecosystems with less thick organic layers (e.g. tropical, Mediterranean forests). On the other hand, forest fires lead to the formation of pyrogenic C, which is biogeochemically stable (Schmidt et al., 2011). Fire-prone systems may thus have particularly large SOC stocks although they lose C in the forest floor during single fires (Eckmeier et al., 2010).

Large reductions in soil C stocks have also been evidenced after stand-replacing windthrows in mountainous forests (Christophel et al., 2015; Mayer et al., 2017b). Post-disturbance C losses were suggested to be caused by decreased organic matter input and increased decomposition rates due to warmer soil conditions following canopy removal. Canopy loss following disturbance can make soils prone to accelerated surface erosion, which may contribute to a reduction in soil C stocks (Berhe et al., 2018; Gerber et al., 2002). Blown-over trees and an associated soil disturbance (e.g. pit-and-mound topography) may also cause soil C losses from deeper subsoil horizons. However, studies also found neutral or even positive effects on soil C stocks following natural forest disturbance (Don et al., 2012; Kobler et al., 2015; Kosunen et al., 2020; Santos et al., 2016). In the following section, selected management activities and disturbance control measures are presented that may mitigate and/or prevent potential negative effects on soil C stocks after natural disturbance.

The intensity of wildfires is often related to the amount of flammable organic material (i.e. branches, twigs, dry plants) on the soil surface. Prescribed burning represents a management activity that reduces fuel loads by planned low-intensity fires to prevent uncontrolled severe wildfires (Alcañiz et al., 2018). Although prescribed burning has been shown to reduce forest floor C stocks, the reductions were smaller (−46%) than after wildfires (−67%) (Nave et al., 2011). The same study showed that prescribed burning did not affect mineral soil C stocks and that the presence of hardwood tree species with less flammable leaves and litter reduced fire impacts on soil C storage. Another strategy to control fire severity may be the establishment of tree-free corridors to reduce fire spread across a forest and to decrease the continuity of fuel loads. Fire suppression often delays but does not prevent forest fires in many cases, as it favors an accumulation of fuel loads that increases the risk of severe wildfires in the long term (Jandl et al., 2007). Following windthrows, insect attacks, and fires, damaged trees are commonly salvage-logged to utilize the timber's merchantable value and to reduce the risk of further pest outbreaks (Leverkus et al., 2018). Although salvage logging is meant to remove dead and dying trees, surviving trees are often also logged, primarily due to more cost-efficient harvesting operations. As additional forest disturbance, salvage logging has been shown to amplify forest floor C losses after windthrow and fire in coniferous forests in North America (Bradford et al., 2012; Kishchuk et al., 2015). Similarly, Hotta et al. (2020) reported lower forest floor C stocks when comparing salvage-logged to non-salvage-logged sites six decades after windthrow in coniferous stands in Japan. Other studies, however, did not find additional effects of salvage logging on soil C stocks following fire and bark-beetle attacks (Avera et al., 2020; Poirier et al., 2014). In some cases, positive effects on soil C storage, either due to higher organic matter input rates from surviving tree- and deadwood litter, decreased decomposition rates due to cooler soil conditions from shading, or lower soil erosion rates were achieved when salvage logging was omitted and dead wood was left in place. As shown in harvesting operations (see *harvesting and harvest residues*), retaining surviving trees and deadwood could be a potential strategy to reduce soil C losses following natural disturbances in certain cases (James et al., 2021; Wan et al., 2018). Deadwood retention, however, may require additional measures (e.g. debarking) to prevent further pest outbreaks in neighboring stands.

Potential negative disturbance effects on soil C stocks may also be mitigated by proactive silvicultural activities that increase stand stability and stand recovery following disturbance. Mixed forests, for example, are less susceptible to damage from bark beetles (Faccoli and Bernardinelli, 2014). A study from North America showed also that the probability of individual tree attack in pine stands was lower at smaller tree diameters and at lower stand densities (Negrón et al., 2008). An uneven-aged stand structure in a Swiss forest was further hypothesized to decrease forest damage by wind (Hanewinkel et al., 2014). Thus, thinned and uneven-aged stands may be more resilient against bark-beetle infestations and windthrows, thereby mitigating net soil C losses (see also *thinning* and *stand density*). Moreover, the promotion of a fast tree regeneration has been suggested to slow post-disturbance soil C losses (Mayer et al., 2017a). Net soil C losses following disturbance may also be reduced if a fast recovery of regenerating trees (either naturally or planted) is fostered.

3 Case study: forest soil carbon storage in Central Europe mountain regions

Mountainous terrains show strong gradients in environmental conditions (climate, parent material, tree species) within a small area, which provide an excellent natural laboratory to identify the relative importance of these factors for SOC storage (Hagedorn et al., 2019). Here, we evaluate the effects of forest types on SOC stocks in Austria and Switzerland with similar topography but different intensity of forest management.

In Austria, the dominating tree species is Norway spruce, both due to its natural dominance in mountain regions and its cultivation at lower elevations. Analyzing the data of the Austrian Forest Soil Survey (Englisch et al., 1992) yielded very small strata for individual deciduous species. Therefore all broadleaf-dominated forests were compared to coniferous forests (Jandl et al., 2021). Overall, the soils of coniferous forests had higher stocks of organic carbon down to 50 cm depth (max. sampling depth). The accumulation of organic C from slowly decomposing tree needles in the forest floor has often been described yet, even in the mineral soil, the SOC stock is higher under conifers than under deciduous trees. The differences in Fig. 3 are statistically insignificant because the tree species effect is overlain by the effects of geology, elevation, and a climatic gradient from a Pannonian climate in the East to a more Atlantic climate in the west of the country.

In Switzerland, forests have primarily been managed as plenter forests with single-tree harvest and promotion of adapted tree species, leading to a near-natural structure of forests. Due to the high costs of wood harvesting, management intensity has declined during the last few decades (Brändli et al., 2020). Consequently, Swiss forests are characterized by a high mean tree age (more than a quarter of Swiss accessible forest area is older than 120 years, see Brändli et al., 2020) and by the greatest stand biomass in Europe (Liski et al.,

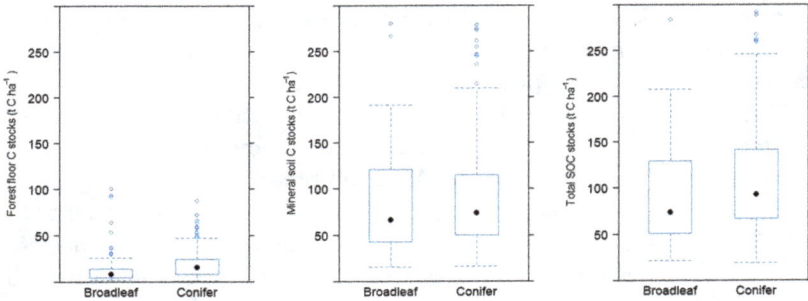

Figure 3 Organic carbon in the soils of Austrian forests. Values are reported to a depth of 50 cm.

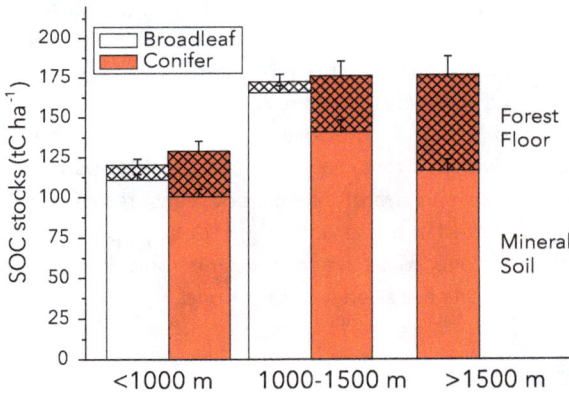

Figure 4 Smaller SOC stocks in the forest floor but greater stocks in the mineral soils in broadleaf than in coniferous forests of Switzerland. Total SOC stocks do not differ between forest types but depend upon elevation (based on approx. 1000 soil profiles).

2002). Swiss forests show a decreasing share of broadleaf trees with increasing elevation due to decreasing temperatures. In their adequate growth region, broadleaf forests dominate (e.g. 68% broadleaf forest in the Swiss plateau).

In Swiss forests, SOC stocks show a general increase with increasing elevation due to less favorable climatic conditions for decomposition (Fig. 4). At a given elevation, total SOC stocks do not differ significantly between the coniferous and broadleaf forests. Nonetheless, the forest types show different vertical SOC distribution with smaller SOC stocks in the forest floor but greater ones in the mineral soil in broadleaf than coniferous forests. This statistical analysis across Swiss forests indicates that forest type is not significantly related to total SOC stocks, while soil chemistry and mineralogy (pH, Fe-, Al-oxides and Ca^{2+}) and precipitation are the most important drivers (Gosheva et al., 2017; Solly et al., 2020). Overall, these findings strongly suggest that, despite the strong effects of tree species on biogeochemical fluxes, vertical SOC distribution and below-ground communities and their functioning, promotion of tree species seems to have only limited potential to influence carbon sequestration.

Favoring deciduous trees over more vulnerable coniferous tree species is expected to increase the stability of forests against emerging climate change-related pressures in Central Europe. Although broadleaf forests have greater SOC stocks in the mineral soil than conifer forests, there is little evidence that total SOC stocks are substantially increased (Grüneberg et al., 2019; Wiesmeier et al., 2013). Forestry follows refined and proven concepts to reach sustainable high-productive forests. High biomass production and optimized forest management have gradually enriched the forest SOC stock during the last few decades. An important factor in Central Europe has been the discontinuation of litter removal for use as animal bedding material (i.e. litter raking) (Führer, 2000). Resorting to other tree species combinations could lead to lower SOC stocks (Prietzel and Bachmann, 2012). However, a wide range of evidence has been provided and the effects are not fully congruent.

Moving toward a higher proportion of deciduous trees also requires auxiliary measures within forest management. Deciduous trees are browsed selectively by roe deer and therefore efficient measures of population control need to be implemented in areas with a high deer population. This argument reflects an old controversy between foresters and hunters. Solutions are discussed but few effects are seen on the ground.

4 Conclusion

Afforestation and reforestation of arable land with low SOC stocks can significantly enhance C sequestration in soils (and biomass) over decades. Yet, it is questionable whether the need to provide food and fuel from agricultural soils for a growing world population is reconcilable with maximized C sequestration in terrestrial ecosystems. A less controversial concept is optimized forest management to maximize stocks of organic carbon sustainably. In our assessment of available data and studies (predominantly temperate forests) we have found several options to increase SOC stocks, but potential effect sizes seem rather limited. Table 2 summarizes the effects of different forest management measures on soil C sequestration and highlights the associated uncertainties and research needs.

Tree species selection can influence C sequestration along the soil profile. Coniferous forests tend to store more C in the topsoil whereas broadleaf forests store more C in the subsoil. However, on the landscape scale (see case study), total soil C stocks are often similar when comparing coniferous and broadleaf forests. Adapting present forests to climate change will entail intermixing of more heat-, drought- and pest-resistant tree species. To what extent such species changes

Table 2 Forest management practices and how they affect forest soil C stocks[a]

Management practice	Effect on soil C	Gaps in knowledge	Research needs
Afforestation and reforestation	Positive on low-C soil (croplands); neutral to negative on high-C soils (grasslands)	Effects on deeper soil C; sequestration pathways (litter input vs. labile C from roots and mycorrhiza)	Long-term studies encompassing several forest rotations; processes involved in soil C buildup (and loss)
Management of tree species	Neutral to positive	Interactions with site preconditions; legacy effects of past tree species; climate change interactions	Moving from common garden experiments to larger scale and integration of legacy effects
Harvest and harvest residues	Negative to neutral	Many soil types are not covered in available studies; nutrient benchmarks for residue removal; long-term effects	Detailed studies on specific harvest techniques and machinery setups; long-term studies in different ecosystem types
Density regulation and thinning	Neutral	Limited information on landscape-scale effects; long-term studies are rare, primarily focusing on temperate and boreal forests	Chronosequence and landscape gradient studies; long-term experiments; inclusion of indirect effects such as disturbance resistance
Fertilization, liming and wood-ash application	Positive for fertilization; unclear for liming and wood-ash application	Long-term effects of liming and wood ash on soil microbiome; long-term effects on deeper soil	Offset emissions from fertilizer production and application; CO_2 emissions from lime; effects on soil biology
Disturbance management	Mostly positive (through avoidance of C loss)	Interaction of climate change and disturbance effects; disturbance-related species shifts	Effects of herbivory and browsing on resilience; integration of remote sensing; storage in charcoal after fire

For each practice, existing knowledge gaps and suggested research to address them are given.

affect soil C sequestration is as yet unclear, since C sequestration not only depends on litter quality and quantity change but also on C-flow below-ground, or the composition of mycorrhizal- and decomposer communities. The potential effects on SOM decomposition (and C sequestration) are correspondingly complex, depending on site-specific pedoclimatic conditions, and the overall effects of a tree species change on soil C sequestration are difficult to predict.

Forests in many eco-regions are already under pressure due to increasing drought, heat waves, and other disturbances. Current forest management therefore often focuses on adaptation (rather than CO_2 mitigation) and the promotion of increased resistance and resilience. Management measures that reduce disturbance risks also reduce the risks of associated soil C losses. For instance, silvicultural practices such as thinning can be utilized to improve stand stability and to adapt tree species composition. Although thinning intensity *per se* seems to have limited effects on soil C sequestration, thinning-related stand stabilization, and structural adaptations may lead to stable SOC stocks in the longer term. Stand-replacing disturbance can impose significant soil C losses, especially if tree regeneration is hampered or delayed. Controlling the density of browsing game populations can be such a management measure in regions where natural predators are extinct.

As a man-made disturbance, tree harvesting operations can lead to transient soil C losses. The magnitude of soil C losses depends on the severity of the harvest intervention, whereby large-scale clear cut with heavy machinery and, for example, whole tree plus stump removal can cause soil C losses that require several years to decades to be balanced out by new C input during stand regeneration. Moreover, it is necessary to align harvesting operations to the prevalent site and soil conditions. On nutrient-rich soils with low C stocks in the forest floor, even whole tree harvest might not affect soil C sequestration, whereas on nutrient-poor soils with generally thick organic horizons, as well as at erosion-prone sites, removal of biomass, and/or harvest residues can substantially reduce soil C sequestration. Physical disturbance, such as soil compaction, can also negatively affect soil C sequestration.

Presently, monetary incentives to optimize tree harvest and forest operations toward soil preservation are rare, and soil C sequestration is usually not the highest priority of forest managers. Harvesting techniques that minimize soil disturbance are available but cost and the high demand for wood and other biomass components make its extraction more attractive than leaving it behind for increased soil C storage. Forest (soil) C accounting and C offset-credit systems might be a way forward, but their implementation will require scientific support since soil C cycling is complex and actual soil C sequestration rates are difficult to quantify.

5   Future trends in research

- What is the carbon sequestration potential of an individual site/soil? It is as yet unclear to what extent soils are already C saturated and thus how strongly current forest soil C stocks can be increased. Specific soil traits (e.g. clay content, Fe and Al-oxides) have been identified as predictors of C sequestration potentials. However, forest management needs more practical guidance; e.g. which soil type at which location and under which forest cover is a promising candidate for C sequestration. A special case is forest soils on carbonate bedrock. The mechanisms in C stabilization in carbonate soils are less well understood than those in acidic soils, and their potential for further C sequestration (or loss) remains uncertain.

- What happens to deep soil C? Effects on soil C storage in deeper subsoil horizons have not been studied in depth, although deep soil C has the longest residence time. Tree species selection can affect rooting depth and thereby alter C inputs into deeper soil. These new C inputs may enhance the decomposition of old existing SOC by priming, leading to few net changes. The rates of C downward leaching can be altered by silvicultural practices (e.g. thinning intensity, harvest method etc.). This could not only trigger deep soil C storage but its decomposition as well. The overall effects are as yet unclear.

- How to ensure sustainable soil C sequestration? How sequestered C is stabilized by the soil matrix will influence its residence time. As an example, afforestation or tree species change from broadleaf to conifer can significantly increase C storage in the forest floor. Forest floor represents a labile C pool that can rapidly decompose again after, for example, a clear-cut harvest or be combusted during fire (slow in fast out). Long-term studies and modeling, including SOC storage in distinct SOC pools, can help determine how current C sequestration measures play out in the longer term.

- Going from case studies to areas: Findings on the sequestration of organic C are often based on case studies with an unclear representativity for regions. There is already an ongoing trend toward regional assessments and landscape gradient approaches. Better integration of remote sensing could provide further opportunities.

- Confining the effects of specific silvicultural activities: while the effects of the top-tier management options have been dealt with (e.g. tree species selection, clear cut or permanent cover, etc.), many specific silvicultural practices are not yet, or only poorly, studied. For instance, it remains unclear if and how different harvesting systems (e.g. cable yarding vs. forwarder vs. tractor winch) affect soil C stocks and sequestration.

- Socioeconomic implications and forest management: there exists only poor understanding of the perception of soil C sequestration within forest management. Specific funding regimes, soil C accounting strategies, as well as monetary incentives for forest soil C sequestration are currently lacking.

6 Where to look for further information

- Jandl, R., Lindner, M., Vesterdal, L., Bauwens, B., Baritz, R., Hagedorn, F., Johnson, D. W., Minkkinen, K. and Byrne, K. A. (2007). How strongly can forest management influence soil carbon sequestration? Geoderma, 137(3-4), 253-268.

- Mayer, M., Prescott, C. E., Abaker, W. E. A., Augusto, L., Cécillon, L., Ferreira, G. W. D., James, J., Jandl, R., Katzensteiner, K., Laclau, J.-P., Laganière, J., Nouvellon, Y., Paré, D., Stanturf, J. A., Vanguelova, E. I. and Vesterdal, L. (2020). Influence of forest management activities on soil organic carbon stocks: A knowledge synthesis. Forest Ecology and Management, 466, 118127. https://doi.org/10.1016/j.foreco.2020.118127.

7 References

Achat, D. L., Fortin, M., Landmann, G., Ringeval, B. and Augusto, L. 2015. Forest soil carbon is threatened by intensive biomass harvesting. *Scientific Reports* 5, 15991.

Alcañiz, M., Outeiro, L., Francos, M. and Úbeda, X. 2018. Effects of prescribed fires on soil properties: a review. *Science of the Total Environment* 613-614, 944-957. https://doi.org/10.1016/j.scitotenv.2017.09.144.

Allen, C. D., Macalady, A. K., Chenchouni, H., Bachelet, D., McDowell, N., Vennetier, M., Kitzberger, T., Rigling, A., Breshears, D. D., Hogg, E. T., Gonzalez, P., Fensham, R., Zhang, Z., Castro, J., Demidova, N., Lim, J., Allard, G., Running, S. W., Semerci, A. and Cobb, N. 2010. A global overview of drought and heat-induced tree mortality reveals emerging climate change risks for forests. *Forest Ecology and Management* 259(4), 660-684.

Ashwood, F., Watts, K., Park, K., Fuentes-Montemayor, E., Benham, S. and Vanguelova, E. I. 2019. Woodland restoration on agricultural land: long-term impacts on soil quality. *Restoration Ecology* 27(6), 1381-1392. https://doi.org/10.1111/rec.13003.

Avera, B. N., Rhoades, C. C., Calderón, F. and Cotrufo, M. F. 2020. Soil C storage following salvage logging and residue management in bark beetle-infested lodgepole pine forests. *Forest Ecology and Management* 472, 118251.

Bárcena, T. G., Gundersen, P. and Vesterdal, L. 2014a. Afforestation effects on SOC in former cropland: oak and spruce chronosequences resampled after 13 years. *Global Change Biology* 20(9), 2938-2952. https://doi.org/10.1111/gcb.12608.

Bárcena, T. G., Kiær, L. P., Vesterdal, L., Stefánsdóttir, H. M., Gundersen, P. and Sigurdsson, B. D. 2014b. Soil carbon stock change following afforestation in northern Europe: a meta-analysis. *Global Change Biology* 20(8), 2393-2405. https://doi.org/10.1111/gcb.12576.

Bastin, J. F., Finegold, Y., Garcia, C., Mollicone, D., Rezende, M., Routh, D., Zohner, C. M. and Crowther, T. W. 2019. The global tree restoration potential. *Science* 365(6448), 76-79.

Bauhus, J., Vor, T., Bartsch, N. and Cowling, A. 2004. The effects of gaps and liming on forest floor decomposition and soil C and N dynamics in a Fagus sylvatica forest. *Canadian Journal of Forest Research* 34(3), 509-518.

Berger, T. W., Swoboda, S., Prohaska, T. and Glatzel, G. 2006. The role of calcium uptake from deep soils for spruce (Picea abies) and beech (Fagus sylvatica). *Forest Ecology and Management* 229(1-3), 234-246.

Berhe, A. A., Barnes, R. T., Six, J. and Marín-Spiotta, E. 2018. Role of soil erosion in biogeochemical cycling of essential elements: carbon, nitrogen, and phosphorus. *Annual Review of Earth and Planetary Sciences* 46(1), 521-548.

Bossio, D. A., Cook-Patton, S. C., Ellis, P. W., Fargione, J., Sanderman, J., Smith, P., Wood, S., Zomer, R. J., von Unger, M., Emmer, I. M. and Griscom, B. W. 2020. The role of soil carbon in natural climate solutions. *Nature Sustainability* 3(5), 391–398. https://doi .org/10.1038/s41893-020-0491-z.

Böttcher, J. and Springob, G. 2001. A carbon balance model for organic layers of acid forest soils. *Journal of Plant Nutrition and Soil Science* 164(4), 399–405.

Bowd, E. J., Banks, S. C., Strong, C. L. and Lindenmayer, D. B. 2019. Long-term impacts of wildfire and logging on forest soils. *Nature Geoscience* 12(2), 113–118.

Bradford, J. B., Fraver, S., Milo, A. M., D'Amato, A. W., Palik, B. and Shinneman, D. J. 2012. Effects of multiple interacting disturbances and salvage logging on forest carbon stocks. *Forest Ecology and Management* 267, 209–214.

Brändli, U., Abegg, M. and Allgaier Leuch, B. 2020. Schweizerisches Landesforstinventar: Ergebnisse der vierten Erhebung 2009-2017 (results of the fourth Swiss National Forest Inventory 2009-2017). Swiss Federal Research Institute for Forest, Snow and Landscape Research, Birmensdorf (ZH) and Federal Office for the Environment (FOEN). Bern.

Bravo-Oviedo, A., Ruiz-Peinado, R., Modrego, P., Alonso, R. and Montero, G. 2015. Forest thinning impact on carbon stock and soil condition in southern European populations of P. sylvestris L. *Forest Ecology and Management* 357, 259–267. https:// doi.org/10.1016/j.foreco.2015.08.005.

Christophel, D., Höllerl, S., Prietzel, J. and Steffens, M. 2015. Long-term development of soil organic carbon and nitrogen stocks after shelterwood- and clear-cutting in a mountain forest in the Bavarian Limestone Alps. *European Journal of Forest Research* 134(4), 623–640. https://doi.org/10.1007/s10342-015-0877-z.

Clarke, N., Kiær, L. P., Janne Kjønaas, O., Bárcena, T. G., Vesterdal, L., Stupak, I., Finér, L., Jacobson, S., Armolaitis, K., Lazdina, D., Stefánsdóttir, H. M. and Sigurdsson, B. D. 2021. Effects of intensive biomass harvesting on forest soils in the Nordic countries and the UK: a meta-analysis. *Forest Ecology and Management* 482, 118877. https:// doi.org/10.1016/j.foreco.2020.118877.

Court, M., van der Heijden, G., Didier, S., Nys, C., Richter, C., Pousse, N., Saint-André, L. and Legout, A. 2018. Long-term effects of forest liming on mineral soil, organic layer and foliage chemistry: insights from multiple beech experimental sites in Northern France. *Forest Ecology and Management* 409, 872–889.

Craig, M. E., Turner, B. L., Liang, C., Clay, K., Johnson, D. J. and Phillips, R. P. 2018. Tree mycorrhizal type predicts within-site variability in the storage and distribution of soil organic matter. *Global Change Biology* 24(8), 3317–3330.

D'Amato, A. W., Bradford, J. B., Fraver, S. and Palik, B. J. 2013. Effects of thinning on drought vulnerability and climate response in north temperate forest ecosystems. *Ecological Applications* 23(8), 1735–1742. https://doi.org/10.1890/13-0677.1.

Dávila Reátegui, H., Poirier, V., Coyea, M. R. and Munson, A. D. 2021. Repeated thinning treatments reduce long-term soil carbon and nitrogen storage: an 87-year study at the Petawawa Research Forest, Canada. *Canadian Journal of Forest Research* 51(2), 190–197.

De Vos, B., Cools, N., Ilvesniemi, H., Vesterdal, L., Vanguelova, E. and Carnicelli, S. 2015. Benchmark values for forest soil carbon stocks in Europe: results from a large scale forest soil survey. *Geoderma* 251–252, 33–46.

De Wandeler, H., Bruelheide, H., Dawud, S. M., Dănilă, G., Domisch, T., Finér, L., Hermy, M., Jaroszewicz, B., Joly, F.-X., Müller, S., Ratcliffe, S., Raulund-Rasmussen, K., Rota, E., Van Meerbeek, K., Vesterdal, L. and Muys, B. 2018. Tree identity rather than tree diversity drives earthworm communities in European forests. *Pedobiologia* 67, 16–25.

Desie, E., Van Meerbeek, K., De Wandeler, H., Bruelheide, H., Domisch, T., Jaroszewicz, B., Joly, F., Vancampenhout, K., Vesterdal, L. and Muys, B. 2020. Positive feedback loop between earthworms, humus form and soil pH reinforces earthworm abundance in European forests. *Functional Ecology* 34(12), 2598-2610.

Dijkstra, F. A., Zhu, B. and Cheng, W. 2021. Root effects on soil organic carbon: a double-edged sword. *New Phytologist* 230(1), 60-65.

Don, A., Bärwolff, M., Kalbitz, K., Andruschkewitsch, R., Jungkunst, H. F. and Schulze, E.-D. 2012. No rapid soil carbon loss after a windthrow event in the High Tatra. *Forest Ecology and Management* 276, 239-246. https://doi.org/10.1016/j.foreco.2012.04.010.

Don, A., Schumacher, J. and Freibauer, A. 2011. Impact of tropical land-use change on soil organic carbon stocks-a meta-analysis. *Global Change Biology* 17(4), 1658-1670.

Eckmeier, E., Egli, M., Schmidt, M. W. I., Schlumpf, N., Nötzli, M., Minikus-Stary, N. and Hagedorn, F. 2010. Preservation of fire-derived carbon compounds and sorptive stabilisation promote the accumulation of organic matter in black soils of the Southern Alps. *Geoderma* 159(1-2), 147-155. https://doi.org/10.1016/j.geoderma.2010.07.006.

Englisch, M., Karrer, G. and Mutsch, F. 1992. Methodische grundlagen. *Mitteilungen der Forstlichen Bundesversuchsanstalt* 168, 5-22.

Faccoli, M. and Bernardinelli, I. 2014. Composition and elevation of spruce forests affect susceptibility to bark beetle attacks: implications for forest management. *Forests* 5(1), 88-102.

Feger, K., Lorenz, K., Raspe, S. and Armbruster, M. 2000. *Mittel-bis langfristige Auswirkungen von Kompensations-bzw. Bodenschutzkalkungen auf die Pedo-und Hydrosphäre. Schlussbericht.-Freiburg.*

Finkral, A. J. and Evans, A. M. 2008. The effects of a thinning treatment on carbon stocks in a northern Arizona ponderosa pine forest. *Forest Ecology and Management* 255(7), 2743-2750. https://doi.org/10.1016/j.foreco.2008.01.041.

Fog, K. 1988. The effect of added nitrogen on the rate of decomposition of organic matter. *Biological Reviews* 63(3), 433-462.

Friedlingstein, P., O'Sullivan, M., Jones, M. W., Andrew, R. M., Hauck, J., Olsen, A., Peters, G. P., Peters, W., Pongratz, J., Sitch, S., Le Quéré, C., Canadell, J. G., Ciais, P., Jackson, R. B., Alin, S., Aragão, L. E. O. C., Arneth, A., Arora, V., Bates, N. R., Becker, M., Benoit-Cattin, A., Bittig, H. C., Bopp, L., Bultan, S., Chandra, N., Chevallier, F., Chini, L. P., Evans, W., Florentie, L., Forster, P. M., Gasser, T., Gehlen, M., Gilfillan, D., Gkritzalis, T., Gregor, L., Gruber, N., Harris, I., Hartung, K., Haverd, V., Houghton, R. A., Ilyina, T., Jain, A. K., Joetzjer, E., Kadono, K., Kato, E., Kitidis, V., Korsbakken, J. I., Landschützer, P., Lefèvre, N., Lenton, A., Lienert, S., Liu, Z., Lombardozzi, D., Marland, G., Metzl, N., Munro, D. R., Nabel, J. E. M. S., Nakaoka, S.-I., Niwa, Y., O'Brien, K., Ono, T., Palmer, P. I., Pierrot, D., Poulter, B., Resplandy, L., Robertson, E., Rödenbeck, C., Schwinger, J., Séférian, R., Skjelvan, I., Smith, A. J. P., Sutton, A. J., Tanhua, T., Tans, P. P., Tian, H., Tilbrook, B., van der Werf, G., Vuichard, N., Walker, A. P., Wanninkhof, R., Watson, A. J., Willis, D., Wiltshire, A. J., Yuan, W., Yue, X. and Zaehle, S. 2020. Global carbon budget 2020. *Earth System Science Data* 12(4), 3269-3340. https://doi.org/10.5194/essd-12-3269-2020.

Friggens, N. L., Hester, A. J., Mitchell, R. J., Parker, T. C., Subke, J. A. and Wookey, P. A. 2020. Tree planting in organic soils does not result in net carbon sequestration on

decadal timescales. *Global Change Biology* 26(9), 5178–5188. https://doi.org/10
.1111/gcb.15229.

Führer, E. 2000. Forest functions, ecosystem stability and management. *Forest Ecology and Management* 132(1), 29–38.

Fulé, P. Z., Waltz, A. E. M., Covington, W. W. and Heinlein, T. A. 2001. Measuring forest restoration effectiveness in reducing hazardous fuels. *Journal of Forestry* 99, 24–29. https://doi.org/10.1093/jof/99.11.24.

Gerber, W., Rickli, C. and Graf, F. 2002. Surface erosion in cleared and uncleared mountain windthrow sites. *Forest Snow and Landscape Research* 77, 109–116.

Gonçalves, A. C. 2020. Thinning: an overview. In: *Silviculture*. IntechOpen. https://www
.intechopen.com/chapters/73015.

Gong, C., Tan, Q., Liu, G. and Xu, M. 2021. Forest thinning increases soil carbon stocks in China. *Forest Ecology and Management* 482, 118812. https://doi.org/10.1016/j
.foreco.2020.118812.

Gosheva, S., Walthert, L., Niklaus, P. A., Zimmermann, S., Gimmi, U. and Hagedorn, F. 2017. Reconstruction of historic forest cover changes indicates minor effects on carbon stocks in Swiss forest soils. *Ecosystems* 20(8), 1512–1528.

Griscom, B. W., Adams, J., Ellis, P. W., Houghton, R. A., Lomax, G., Miteva, D. A., Schlesinger, W. H., Shoch, D., Siikamäki, J. V., Smith, P., Woodbury, P., Zganjar, C., Blackman, A., Campari, J., Conant, R. T., Delgado, C., Elias, P., Gopalakrishna, T., Hamsik, M. R., Herrero, M., Kiesecker, J., Landis, E., Laestadius, L., Leavitt, S. M., Minnemeyer, S., Polasky, S., Potapov, P., Putz, F. E., Sanderman, J., Silvius, M., Wollenberg, E. and Fargione, J. 2017. Natural climate solutions. *Proceedings of the National Academy of Sciences of the United States of America* 114(44), 11645–11650.

Grüneberg, E., Schöning, I., Riek, W., Ziche, D. and Evers, J. 2019. Carbon stocks and carbon stock changes in German Forest Soils. In: ⬚. Wellbrock ⬚nd ⬚. Bolte (eds) *Status and Dynamics of Forests in Germany. Ecological Studies, vol 237*. Springer, Cham, 167–198. https://doi.org/10.1007/978-3-030-15734-0_6.

Grüneberg, E., Ziche, D. and Wellbrock, N. 2014. Organic carbon stocks and sequestration rates of forest soils in Germany. *Global Change Biology* 20(8), 2644–2662.

Hagedorn, F., Gavazov, K. and Alexander, J. M. 2019. Above-and belowground linkages shape responses of mountain vegetation to climate change. *Science* 365(6458), 1119–1123.

Hagedorn, F., Spinnler, D. and Siegwolf, R. 2003. Increased N deposition retards mineralization of old soil organic matter. *Soil Biology and Biochemistry* 35(12), 1683–1692.

Hanewinkel, M., Kuhn, T., Bugmann, H., Lanz, A. and Brang, P. 2014. Vulnerability of uneven-aged forests to storm damage. *Forestry* 87(4), 525–534.

Hansson, K., Kleja, D. B., Kalbitz, K. and Larsson, H. 2010. Amounts of carbon mineralised and leached as DOC during decomposition of Norway spruce needles and fine roots. *Soil Biology and Biochemistry* 42(2), 178–185. https://doi.org/10.1016/j
.soilbio.2009.10.013.

Harris, N. L., Gibbs, D. A., Baccini, A., Birdsey, R. A., de Bruin, S., Farina, M., Fatoyinbo, L., Hansen, M. C., Herold, M., Houghton, R. A., Potapov, P. V., Suarez, D. R., Roman-Cuesta, R. M., Saatchi, S. S., Slay, C. M., Turubanova, S. A. and Tyukavina, A. 2021. Global maps of twenty-first century forest carbon fluxes. *Nature Climate Change* 11(3), 234–240. https://doi.org/10.1038/s41558-020-00976-6.

Heděnec, P., Nilsson, L. O., Zheng, H., Gundersen, P., Schmidt, I. K., Rousk, J. and Vesterdal, L. 2020. Mycorrhizal association of common European tree species shapes biomass and metabolic activity of bacterial and fungal communities in soil. *Soil Biology and Biochemistry* 149, 107933. https://doi.org/10.1016/j.soilbio.2020.107933.

Hildebrand, E. 1996. Warum müssen wir Waldböden kalken. *Agrarforschung in Baden-Württemberg* 26, 53–65.

Hiltbrunner, D., Zimmermann, S. and Hagedorn, F. 2013. Afforestation with Norway spruce on a subalpine pasture alters carbon dynamics but only moderately affects soil carbon storage. *Biogeochemistry* 115(1–3), 251–266.

Hlásny, T., Zimová, S., Merganičová, K., Štěpánek, P., Modlinger, R. and Turčáni, M. 2021. Devastating outbreak of bark beetles in the Czech Republic: drivers, impacts, and management implications. *Forest Ecology and Management* 490, 119075.

Hong, S., Yin, G., Piao, S., Dybzinski, R., Cong, N., Li, X., Wang, K., Peñuelas, J., Zeng, H. and Chen, A. 2020. Divergent responses of soil organic carbon to afforestation. *Nature Sustainability* 3(9), 694–700.

Hoover, C. M. 2011. Management impacts on forest floor and soil organic carbon in northern temperate forests of the US. *Carbon Balance and Management* 6, 1–8.

Hotta, W., Morimoto, J., Inoue, T., Suzuki, S. N., Umebayashi, T., Owari, T., Shibata, H., Ishibashi, S., Hara, T. and Nakamura, F. 2020. Recovery and allocation of carbon stocks in boreal forests 64 years after catastrophic windthrow and salvage logging in northern Japan. *Forest Ecology and Management* 468, 118169.

Hume, A. M., Chen, H. Y. H. and Taylor, A. R. 2018. Intensive forest harvesting increases susceptibility of northern forest soils to carbon, nitrogen and phosphorus loss. *Journal of Applied Ecology* 55(1), 246–255.

Huotari, N., Tillman-Sutela, E., Moilanen, M. and Laiho, R. 2015. Recycling of ash–For the good of the environment? *Forest Ecology and Management* 348, 226–240.

Hyvönen, R., Kaarakka, L., Leppälammi-Kujansuu, J., Olsson, B. A., Palviainen, M., Vegerfors-Persson, B. and Helmisaari, H.-S. 2016. Effects of stump harvesting on soil C and N stocks and vegetation 8–13 years after clear-cutting. *Forest Ecology and Management* 371, 23–32.

Jackson, R. B., Jobbágy, E. G., Avissar, R., Roy, S. B., Barrett, D. J., Cook, C. W., Farley, K. A., Maitre, D. C. le, McCarl, B. A. and Murray, B. C. 2005. Trading water for carbon with biological carbon sequestration. *Science* 310(5756), 1944–1947. https://doi.org/10.1126/science.1119282.

James, J. and Harrison, R. 2016. The effect of harvest on forest soil carbon: a meta-analysis. *Forests* 7(12), 308. https://doi.org/10.3390/f7120308.

James, J., Page-Dumroese, D., Busse, M., Palik, B., Zhang, J., Eaton, B., Slesak, R., Tirocke, J. and Kwon, H. 2021. Effects of forest harvesting and biomass removal on soil carbon and nitrogen: two complementary meta-analyses. *Forest Ecology and Management* 485, 118935.

Jandl, R. 2020. Climate-induced challenges of Norway spruce in Northern Austria. *Trees, Forests and People* 1, 100008. https://doi.org/10.1016/j.tfp.2020.100008.

Jandl, R., Ledermann, T., Kindermann, G. and Weiss, P. 2021. Soil organic carbon stocks in mixed-deciduous and coniferous forests in Austria. *Frontiers in Forests and Global Change* 4, 69.

Jandl, R., Lindner, M., Vesterdal, L., Bauwens, B., Baritz, R., Hagedorn, F., Johnson, D. W., Minkkinen, K. and Byrne, K. A. 2007. How strongly can forest management influence soil carbon sequestration? *Geoderma* 137(3–4), 253–268.

Johnson, D. W. and Curtis, P. S. 2001. Effects of forest management on soil C and N storage: meta analysis. *Forest Ecology and Management* 140(2-3), 227-238.

Jonard, M., Nicolas, M., Coomes, D. A., Caignet, I., Saenger, A. and Ponette, Q. 2017. Forest soils in France are sequestering substantial amounts of carbon. *Science of the Total Environment* 574, 616-628.

Jovani-Sancho, A. J., Cummins, T. and Byrne, K. A. 2021. Soil carbon balance of afforested peatlands in the maritime temperate climatic zone. *Global Change Biology* 27(15), 3681-3698. https://doi.org/10.1111/gcb.15654.

Jurevics, A., Peichl, M., Olsson, B. A., Strömgren, M. and Egnell, G. 2016. Slash and stump harvest have no general impact on soil and tree biomass C pools after 32-39 years. *Forest Ecology and Management* 371, 33-41.

Jurgensen, M., Tarpey, R., Pickens, J., Kolka, R. and Palik, B. 2012. Long-term effect of silvicultural thinnings on soil carbon and nitrogen pools. *Soil Science Society of America Journal* 76(4), 1418-1425.

Kaarakka, L., Hyvönen, R., Strömgren, M., Palviainen, M., Persson, T., Olsson, B. A., Launonen, E., Vegerfors, B. and Helmisaari, H.-S. 2016. Carbon and nitrogen pools and mineralization rates in boreal forest soil after stump harvesting. *Forest Ecology and Management* 377, 61-70.

Kaiser, K. and Guggenberger, G. 2003. Mineral surfaces and soil organic matter. *European Journal of Soil Science* 54(2), 219-236.

Kim, S., Kim, C., Han, S. H., Lee, S.-T. and Son, Y. 2018. A multi-site approach toward assessing the effect of thinning on soil carbon contents across temperate pine, oak, and larch forests. *Forest Ecology and Management* 424, 62-70. https://doi.org/10.1016/j.foreco.2018.04.040.

Kishchuk, B. E., Thiffault, E., Lorente, M., Quideau, S., Keddy, T. and Sidders, D. 2015. Decadal soil and stand response to fire, harvest, and salvage-logging disturbances in the western boreal mixedwood forest of Alberta, Canada. *Canadian Journal of Forest Research* 45(2), 141-152.

Kjøller, R. and Clemmensen, K. E. 2008. *The Impact of Liming on Ectomycorrhizal Fungal Communities in Coniferous Forests in Southern Sweden*. Skogsstyrelsens förlag.

Klein, D., Fuentes, J. P., Schmidt, A., Schmidt, H. and Schulte, A. 2008. Soil organic C as affected by silvicultural and exploitative interventions in Nothofagus pumilio forests of the Chilean Patagonia. *Forest Ecology and Management* 255(10), 3549-3555.

Kobler, J., Jandl, R., Dirnböck, T., Mirtl, M. and Schindlbacher, A. 2015. Effects of stand patchiness due to windthrow and bark beetle abatement measures on soil CO_2 efflux and net ecosystem productivity of a managed temperate mountain forest. *European Journal of Forest Research* 134(4), 683-692. https://doi.org/10.1007/s10342-015-0882-2.

Kosunen, M., Peltoniemi, K., Pennanen, T., Lyytikäinen-Saarenmaa, P., Adamczyk, B., Fritze, H., Zhou, X. and Starr, M. 2020. Storm and Ips typographus disturbance effects on carbon stocks, humus layer carbon fractions and microbial community composition in boreal Picea abies stands. *Soil Biology and Biochemistry* 148, 107853.

Kreutzer, K. 1995. Effects of forest liming on soil processes. *Plant and Soil* 168-169(1), 447-470.

Law, B. E. and Waring, R. H. 2015. Carbon implications of current and future effects of drought, fire and management on Pacific Northwest forests. *Forest Ecology and Management* 355, 4-14. https://doi.org/10.1016/j.foreco.2014.11.023.

Leverkus, A. B., Rey Benayas, J. M., Castro, J., Boucher, D., Brewer, S., Collins, B. M., Donato, D., Fraver, S., Kishchuk, B. E., Lee, E.-J., Lindenmayer, D. B., Lingua, E., Macdonald, E., Marzano, R., Rhoades, C. C., Royo, A., Thorn, S., Wagenbrenner, J. W., Waldron, K., Wohlgemuth, T. and Gustafsson, L. 2018. Salvage logging effects on regulating and supporting ecosystem services—a systematic map. *Canadian Journal of Forest Research* 48(9), 983–1000.

Li, R., Guan, X., Han, J., Zhang, Y., Zhang, W., Wang, J., Huang, Y., Xu, M., Chen, L., Wang, S. and Yang, Q. 2021. Litter decomposition was retarded by understory removal but was unaffected by thinning in a Chinese fir [Cunninghamia lanceolata (Lamb.) Hook] plantation. *Applied Soil Ecology* 163, 103968. https://doi.org/10.1016/j.apsoil.2021 .103968.

Lim, H., Olsson, B. A., Lundmark, T., Dahl, J. and Nordin, A. 2020. Effects of whole-tree harvesting at thinning and subsequent compensatory nutrient additions on carbon sequestration and soil acidification in a boreal forest. *GCB Bioenergy* 12(11), 992–1001.

Liski, J., Perruchoud, D. and Karjalainen, T. 2002. Increasing carbon stocks in the forest soils of Western Europe. *Forest Ecology and Management* 169(1-2), 159–175.

Marschner, B. and Wilczynski, A. W. 1991. The effect of liming on quantity and chemical composition of soil organic matter in a pine forest in Berlin, Germany. *Plant and Soil* 137(2), 229–236.

Matzner, E., Khanna, P. K., Meiwes, K. J. and Ulrich, B. 1985. Effects of fertilization and liming on the chemical soil conditions and element distribution in forest soils. *Plant and Soil* 87(3), 405–415. https://doi.org/10.1007/BF02181907.

Mayer, M., Matthews, B., Rosinger, C., Sandén, H., Godbold, D. L. and Katzensteiner, K. 2017a. Tree regeneration retards decomposition in a temperate mountain soil after forest gap disturbance. *Soil Biology and Biochemistry* 115, 490–498.

Mayer, M., Sandén, H., Rewald, B., Godbold, D. L. and Katzensteiner, K. 2017b. Increase in heterotrophic soil respiration by temperature drives decline in soil organic carbon stocks after forest windthrow in a mountainous ecosystem. *Functional Ecology* 31(5), 1163–1172.

Mayer, M., Pousse, N. and James, J. 2021. Harvest systems that limit soil disturbance and reduced impact logging. In: *Recarbonizing Global Soils: A Technical Manual of Recommended Management Practices*. Food and Agriculture Organization of the United Nations, Rome, Italy.

Mayer, M., Prescott, C. E., Abaker, W. E. A., Augusto, L., Cécillon, L., Ferreira, G. W. D., James, J., Jandl, R., Katzensteiner, K., Laclau, J.-P., Laganière, J., Nouvellon, Y., Paré, D., Stanturf, J. A., Vanguelova, E. I. and Vesterdal, L. 2020. Tamm review: Influence of forest management activities on soil organic carbon stocks: a knowledge synthesis. *Forest Ecology and Management* 466, 118127. https://doi.org/10.1016/j.foreco .2020.118127.

Melillo, J. M., Frey, S. D., DeAngelis, K. M., Werner, W. J., Bernard, M. J., Bowles, F. P., Pold, G., Knorr, M. A. and Grandy, A. S. 2017. Long-term pattern and magnitude of soil carbon feedback to the climate system in a warming world. *Science* 358(6359), 101–105.

Minasny, B., Malone, B. P., McBratney, A. B., Angers, D. A., Arrouays, D., Chambers, A., Chaplot, V., Chen, Z.-S., Cheng, K., Das, B. S., Field, D. J., Gimona, A., Hedley, C. B., Hong, S. Y., Mandal, B., Marchant, B. P., Martin, M., McConkey, B. G., Mulder, V. L., O'Rourke, S., Richer-de-Forges, A. C., Odeh, I., Padarian, J., Paustian, K., Pan, G.,

Poggio, L., Savin, I., Stolbovoy, V., Stockmann, U., Sulaeman, Y., Tsui, C., Vågen, T., van Wesemael, B. and Winowiecki, L. 2017. Soil carbon 4 per mille. *Geoderma* 292, 59–86.

Mobley, M. L., Lajtha, K., Kramer, M. G., Bacon, A. R., Heine, P. R. and Richter, D. D. 2015. Surficial gains and subsoil losses of soil carbon and nitrogen during secondary forest development. *Global Change Biology* 21(2), 986–996. https://doi.org/10.1111/gcb.12715.

Na, M., Sun, X., Zhang, Y., Sun, Z. and Rousk, J. 2021. Higher stand densities can promote soil carbon storage after conversion of temperate mixed natural forests to larch plantations. *European Journal of Forest Research* 140(2), 373–386. https://doi.org/10.1007/s10342-020-01346-9.

Nave, L. E., Vance, E. D., Swanston, C. W. and Curtis, P. S. 2009. Impacts of elevated N inputs on north temperate forest soil C storage, C/N, and net N-mineralization. *Geoderma* 153(1–2), 231–240.

Nave, L. E., Domke, G. M., Hofmeister, K. L., Mishra, U., Perry, C. H., Walters, B. F. and Swanston, C. W. 2018. Reforestation can sequester two petagrams of carbon in US topsoils in a century. *Proceedings of the National Academy of Sciences of the United States of America* 115(11), 2776–2781.

Nave, L. E., Vance, E. D., Swanston, C. W. and Curtis, P. S. 2011. Fire effects on temperate forest soil C and N storage. *Ecological Applications* 21(4), 1189–1201.

Nave, L. E., Vance, E. D., Swanston, C. W. and Curtis, P. S. 2010. Harvest impacts on soil carbon storage in temperate forests. *Forest Ecology and Management* 259(5), 857–866. https://doi.org/10.1016/j.foreco.2009.12.009.

Negrón, J. F., Allen, K., Cook, B. and Withrow Jr., J. R. 2008. Susceptibility of ponderosa pine, Pinus ponderosa (Dougl. ex Laws.), to mountain pine beetle, Dendroctonus ponderosae Hopkins, attack in uneven-aged stands in the Black Hills of South Dakota and Wyoming USA. *Forest Ecology and Management* 254(2), 327–334.

Nilsen, P. and Strand, L. T. 2008. Thinning intensity effects on carbon and nitrogen stores and fluxes in a Norway spruce (Picea abies (L.) Karst.) stand after 33 years. *Forest Ecology and Management* 256(3), 201–208.

Noormets, A. and Nouvellon, Y. 2015. Introduction for special issue: carbon, water and nutrient cycling in managed forests. *Forest Ecology and Management* 355, 1–3. https://doi.org/10.1016/j.foreco.2015.08.022.

Pan, Y., Birdsey, R. A., Fang, J., Houghton, R., Kauppi, P. E., Kurz, W. A., Phillips, O. L., Shvidenko, A., Lewis, S. L., Canadell, J. G., Ciais, P., Jackson, R. B., Pacala, S. W., McGuire, A. D., Piao, S., Rautiainen, A., Sitch, S. and Hayes, D. 2011. A large and persistent carbon sink in the world's forests. *Science* 333(6045), 988–993. https://doi.org/10.1126/science.1201609.

Peng, Y., Schmidt, I. K., Zheng, H., Heděnec, P., Bachega, L. R., Yue, K., Wu, F. and Vesterdal, L. 2020. Tree species effects on topsoil carbon stock and concentration are mediated by tree species type, mycorrhizal association, and N-fixing ability at the global scale. *Forest Ecology and Management* 478, 118510.

Pérez-Izquierdo, L., Clemmensen, K. E., Strengbom, J., Granath, G., Wardle, D. A., Nilsson, M. and Lindahl, B. D. 2021. Crown-fire severity is more important than ground-fire severity in determining soil fungal community development in the boreal forest. *Journal of Ecology* 109(1), 504–518.

Persson, T., Rudebeck, A. and Wirén, A. 1995. Pools and fluxes of carbon and nitrogen in 40-year-old forest liming experiments in southern Sweden. *Water, Air, and Soil Pollution* 85(2), 901–906.

Poeplau, C. and Don, A. 2013. Sensitivity of soil organic carbon stocks and fractions to different land-use changes across Europe. *Geoderma* 192, 189–201. https://doi.org /10.1016/j.geoderma.2012.08.003.

Poirier, V., Paré, D., Boiffin, J. and Munson, A. D. 2014. Combined influence of fire and salvage logging on carbon and nitrogen storage in boreal forest soil profiles. *Forest Ecology and Management* 326, 133–141.

Popp, J., Lakner, Z., Harangi-Rakos, M. and Fari, M. 2014. The effect of bioenergy expansion: food, energy, and environment. *Renewable and Sustainable Energy Reviews* 32, 559–578.

Powers, M., Kolka, R., Palik, B., McDonald, R. and Jurgensen, M. 2011. Long-term management impacts on carbon storage in Lake States forests. *Forest Ecology and Management* 262(3), 424–431. https://doi.org/10.1016/j.foreco.2011.04.008.

Powers, M. D., Kolka, R. K., Bradford, J. B., Palik, B. J., Fraver, S. and Jurgensen, M. F. 2012. Carbon stocks across a chronosequence of thinned and unmanaged red pine (Pinus resinosa) stands. *Ecological Applications* 22(4), 1297–1307. https://doi.org/10.1890 /11-0411.1.

Prietzel, J. and Bachmann, S. 2012. Changes in soil organic C and N stocks after forest transformation from Norway spruce and Scots pine into Douglas fir, Douglas fir/ spruce, or European beech stands at different sites in Southern Germany. *Forest Ecology and Management* 269, 134–148. https://doi.org/10.1016/j.foreco.2011.12 .034.

Prietzel, J., Zimmermann, L., Schubert, A. and Christophel, D. 2016. Organic matter losses in German Alps forest soils since the 1970s most likely caused by warming. *Nature Geoscience* 9(7), 543–548.

Puhlick, J. J., Fernandez, I. J. and Weiskittel, A. R. 2016. Evaluation of forest management effects on the mineral soil carbon pool of a lowland, mixed-species forest in Maine, USA. *Canadian Journal of Soil Science* 96(2), 207–218.

Rabot, E., Wiesmeier, M., Schlüter, S. and Vogel, H.-J. 2018. Soil structure as an indicator of soil functions: a review. *Geoderma* 314, 122–137. https://doi.org/10.1016/j .geoderma.2017.11.009.

Rasse, D. P., Rumpel, C. and Dignac, M.-F. 2005. Is soil carbon mostly root carbon? Mechanisms for a specific stabilisation. *Plant and Soil* 269(1–2), 341–356. https://doi .org/10.1007/s11104-004-0907-y.

Reich, P. B., Oleksyn, J., Modrzynski, J., Mrozinski, P., Hobbie, S. E., Eissenstat, D. M., Chorover, J., Chadwick, O. A., Hale, C. M. and Tjoelker, M. G. 2005. Linking litter calcium, earthworms and soil properties: a common garden test with 14 tree species. *Ecology Letters* 8(8), 811–818.

Rosenberg, O., Persson, T., Högbom, L. and Jacobson, S. 2010. Effects of wood-ash application on potential carbon and nitrogen mineralisation at two forest sites with different tree species, climate and N status. *Forest Ecology and Management* 260(4), 511–518.

Ruiz-Peinado, R., Bravo-Oviedo, A., Montero, G. and Del Río, M. 2016. Carbon stocks in a Scots pine afforestation under different thinning intensities management. *Mitigation and Adaptation Strategies for Global Change* 21, 1059–1072.

Santos, L. T. dos, Magnabosco Marra, D., Trumbore, S., Camargo, P. B. de, Negrón-Juárez, R. I., Lima, A. J. N., Ribeiro, G. H. P. M., Santos, J. dos and Higuchi, N. 2016. Windthrows increase soil carbon stocks in a central Amazon forest. *Biogeosciences* 13(4), 1299–1308.

Santos, R. S., Oliveira, F. C. C., Ferreira, G. W. D., Ferreira, M. A., Araujo, E. F. and Silva, I. R. 2020. Carbon and nitrogen dynamics in soil organic matter fractions following eucalypt afforestation in southern Brazilian grasslands (Pampas). *Agriculture, Ecosystems and Environment* 301, 106979.

Schäffer, J., Geißen, V., Hoch, R. and von Wilpert, K. 2001. Waldkalkung belebt Böden wieder. *AFZ-Der Wald* Bd 21, 1106-1109.

Scharlemann, J. P., Tanner, E. V., Hiederer, R. and Kapos, V. 2014. Global soil carbon: understanding and managing the largest terrestrial carbon pool. *Carbon Management* 5(1), 81-91. https://doi.org/10.4155/cmt.13.77.

Schindlbacher, A., Schnecker, J., Takriti, M., Borken, W. and Wanek, W. 2015. Microbial physiology and soil CO_2 efflux after 9 years of soil warming in a temperate forest – no indications for thermal adaptations. *Global Change Biology* 21(11), 4265-4277. https://doi.org/10.1111/gcb.12996.

Schlesinger, W. H. 1990. Evidence from chronosequence studies for a low carbon-storage potential of soils. *Nature* 348(6298), 232-234. https://doi.org/10.1038/348232a0.

Schmidt, M. W. I., Torn, M. S., Abiven, S., Dittmar, T., Guggenberger, G., Janssens, I. A., Kleber, M., Kögel-Knabner, I., Lehmann, J., Manning, D. A. C., Nannipieri, P., Rasse, D. P., Weiner, S. and Trumbore, S. E. 2011. Persistence of soil organic matter as an ecosystem property. *Nature* 478(7367), 49-56. https://doi.org/10.1038/nature10386.

Seidl, R., Schelhaas, M. J., Rammer, W. and Verkerk, P. J. 2014. Increasing forest disturbances in Europe and their impact on carbon storage. *Nature Climate Change* 4(9), 806-810.

Seidl, R., Thom, D., Kautz, M., Martin-Benito, D., Peltoniemi, M., Vacchiano, G., Wild, J., Ascoli, D., Petr, M., Honkaniemi, J., Lexer, M. J., Trotsiuk, V., Mairota, P., Svoboda, M., Fabrika, M., Nagel, T. A. and Reyer, C. P. O. 2017. Forest disturbances under climate change. *Nature Climate Change* 7, 395-402. https://doi.org/10.1038/nclimate3303.

Senf, C., Pflugmacher, D., Zhiqiang, Y., Sebald, J., Knorn, J., Neumann, M., Hostert, P. and Seidl, R. 2018. Canopy mortality has doubled in Europe's temperate forests over the last three decades. *Nature Communications* 9(1), 4978. https://doi.org/10.1038/s41467-018-07539-6.

Silver, W. L., Ostertag, R. and Lugo, A. E. 2000. The potential for carbon sequestration Through reforestation of abandoned tropical agricultural and pasture lands. *Restoration Ecology* 8(4), 394-407. https://doi.org/10.1046/j.1526-100x.2000.80054.x.

Simola, H., Pitkänen, A. and Turunen, J. 2012. Carbon loss in drained forestry peatlands in Finland, estimated by re-sampling peatlands surveyed in the 1980s. *European Journal of Soil Science* 63(6), 798-807. https://doi.org/10.1111/j.1365-2389.2012.01499.x.

Smith, P., Clark, H., Dong, H., Elsiddig, E. A., Haberl, H., Harper, R., House, J., Jafari, M., et al. 2014. Chapter 11 - Agriculture, forestry and other land use (AFOLU). In: *Climate Change 2014: Mitigation of Climate Change. IPCC Working Group III Contribution to AR5*. Cambridge University Press. https://pure.iiasa.ac.at/id/eprint/11115/.

Smith, P., Davis, S. J., Creutzig, F., Fuss, S., Minx, J., Gabrielle, B., Kato, E., Jackson, R. B., Cowie, A., Kriegler, E., van Vuuren, D. P., Rogelj, J., Ciais, P., Milne, J., Canadell, J. G., McCollum, D., Peters, G., Andrew, R., Krey, V., Shrestha, G., Friedlingstein, P., Gasser, T., Grübler, A., Heidug, W. K., Jonas, M., Jones, C. D., Kraxner, F., Littleton, E., Lowe, J., Moreira, J. R., Nakicenovic, N., Obersteiner, M., Patwardhan, A., Rogner, M., Rubin, E.,

Sharifi, A., Torvanger, A., Yamagata, Y., Edmonds, J. and Yongsung, C. 2016. Biophysical and economic limits to negative CO_2 emissions. *Nature Climate Change* 6(1), 42–50.

Sohn, J. A., Saha, S. and Bauhus, J. 2016. Potential of forest thinning to mitigate drought stress: a meta-analysis. *Forest Ecology and Management* 380, 261–273. https://doi.org/10.1016/j.foreco.2016.07.046.

Solly, E. F., Weber, V., Zimmermann, S., Walthert, L., Hagedorn, F. and Schmidt, M. W. I. 2020. A critical evaluation of the relationship between the effective cation exchange capacity and soil organic carbon content in Swiss forest soils. *Frontiers in Forests and Global Change* 3, 98.

Stanturf, J. A., Kleine, M., Mansourian, S., Parrotta, J., Madsen, P., Kant, P., Burns, J. and Bolte, A. 2019. Implementing forest landscape restoration under the Bonn Challenge: a systematic approach. *Annals of Forest Science* 76(2), 50. https://doi.org/10.1007/s13595-019-0833-z.

Stewart, C. E., Paustian, K., Conant, R. T., Plante, A. F. and Six, J. 2007. Soil carbon saturation: concept, evidence and evaluation. *Biogeochemistry* 86(1), 19–31.

Strömgren, M., Egnell, G. and Olsson, B. A. 2013. Carbon stocks in four forest stands in Sweden 25 years after harvesting of slash and stumps. *Forest Ecology and Management* 290, 59–66.

Sun, O. J., Campbell, J., Law, B. E. and Wolf, V. 2004. Dynamics of carbon stocks in soils and detritus across chronosequences of different forest types in the Pacific Northwest, USA. *Global Change Biology* 10(9), 1470–1481.

Tang, J., Bolstad, P. V. and Martin, J. G. 2009. Soil carbon fluxes and stocks in a Great Lakes forest chronosequence. *Global Change Biology* 15(1), 145–155.

Temperton, V. M., Buchmann, N., Buisson, E., Durigan, G., Kazmierczak, L., Perring, M. P., de Sá Dechoum, M., Veldman, J. W. and Overbeck, G. E. 2019. Step back from the forest and step up to the Bonn Challenge: how a broad ecological perspective can promote successful landscape restoration. *Restoration Ecology* 27, 705–719.

Thom, D. and Seidl, R. 2016. Natural disturbance impacts on ecosystem services and biodiversity in temperate and boreal forests. *Biological Reviews of the Cambridge Philosophical Society* 91(3), 760–781. https://doi.org/10.1111/brv.12193.

van der Linde, S., Suz, L. M., Orme, C. D. L., Cox, F., Andreae, H., Asi, E., Atkinson, B., Benham, S., Carroll, C., Cools, N., De Vos, B., Dietrich, H. P., Eichhorn, J., Gehrmann, J., Grebenc, T., Gweon, H. S., Hansen, K., Jacob, F., Kristöfel, F., Lech, P., Manninger, M., Martin, J., Meesenburg, H., Merilä, P., Nicolas, M., Pavlenda, P., Rautio, P., Schaub, M., Schröck, H. W., Seidling, W., Šrámek, V., Thimonier, A., Thomsen, I. M., Titeux, H., Vanguelova, E., Verstraeten, A., Vesterdal, L., Waldner, P., Wijk, S., Zhang, Y., Žlindra, D. and Bidartondo, M. I. 2018. Environment and host as large-scale controls of ectomycorrhizal fungi. *Nature* 558(7709), 243–248. https://doi.org/10.1038/s41586-018-0189-9.

Vanguelova, E., Pitman, R., Luiro, J. and Helmisaari, H.-S. 2010. Long term effects of whole tree harvesting on soil carbon and nutrient sustainability in the UK. *Biogeochemistry* 101(1–3), 43–59.

Vanguelova, E. I., Crow, P., Benham, S., Pitman, R., Forster, J., Eaton, E. L. and Morison, J. I. L. 2019. Impact of Sitka spruce (Picea sitchensis (Bong.) Carr.) afforestation on the carbon stocks of peaty gley soils – a chronosequence study in the north of England. *Forestry* 92(3), 242–252. https://doi.org/10.1093/forestry/cpz013.

Veldman, J. W., Aleman, J. C., Alvarado, S. T., Anderson, T. M., Archibald, S., Bond, W. J., Boutton, T. W., Buchmann, N., Buisson, E., Canadell, J. G., Dechoum, M. S.,

Diaz-Toribio, M. H., Durigan, G., Ewel, J. J., Fernandes, G. W., Fidelis, A., Fleischman, F., Good, S. P., Griffith, D. M., Hermann, J. M., Hoffmann, W. A., Le Stradic, S., Lehmann, C. E. R., Mahy, G., Nerlekar, A. N., Nippert, J. B., Noss, R. F., Osborne, C. P., Overbeck, G. E., Parr, C. L., Pausas, J. G., Pennington, R. T., Perring, M. P., Putz, F. E., Ratnam, J., Sankaran, M., Schmidt, I. B., Schmitt, C. B., Silveira, F. A. O., Staver, A. C., Stevens, N., Still, C. J., Strömberg, C. A. E., Temperton, V. M., Varner, J. M. and Zaloumis, N. P. 2019. Comment on "The global tree restoration potential." *Science* 366(6463). https://doi.org/10.1126/science.aay7976.

Vesterdal, L., Clarke, N., Sigurdsson, B. D. and Gundersen, P. 2013. Do tree species influence soil carbon stocks in temperate and boreal forests? *Forest Ecology and Management* 309, 4–18.

Vesterdal, L., Dalsgaard, M., Felby, C., Raulund-Rasmussen, K. and Jørgensen, B. B. 1995. Effects of thinning and soil properties on accumulation of carbon, nitrogen and phosphorus in the forest floor of Norway spruce stands. *Forest Ecology and Management* 77(1–3), 1–10. https://doi.org/10.1016/0378-1127(95)03579-Y.

Vitasse, Y., Bottero, A., Rebetez, M., Conedera, M., Augustin, S., Brang, P. and Tinner, W. 2019. What is the potential of silver fir to thrive under warmer and drier climate? *European Journal of Forest Research* 138(4), 547–560.

Wan, X., Xiao, L., Vadeboncoeur, M. A., Johnson, C. E. and Huang, Z. 2018. Response of mineral soil carbon storage to harvest residue retention depends on soil texture: a meta-analysis. *Forest Ecology and Management* 408, 9–15. https://doi.org/10.1016/j.foreco.2017.10.028.

Wang, Q., Zhong, M. and Wang, S. 2012. A meta-analysis on the response of microbial biomass, dissolved organic matter, respiration, and N mineralization in mineral soil to fire in forest ecosystems. *Forest Ecology and Management* 271, 91–97. https://doi.org/10.1016/j.foreco.2012.02.006.

Warren, K. L. and Ashton, M. S. 2014. Change in soil and forest floor carbon after shelterwood harvests in a New England Oak-Hardwood Forest, USA. *International Journal of Forestry Research* 2014, 1–9.

Wiesmeier, M., Prietzel, J., Barthold, F., Spörlein, P., Geuß, U., Hangen, E., Reischl, A., Schilling, B., von Lützow, M. and Kögel-Knabner, I. 2013. Storage and drivers of organic carbon in forest soils of southeast Germany (Bavaria)–Implications for carbon sequestration. *Forest Ecology and Management* 295, 162–172.

Wiesmeier, M., Urbanski, L., Hobley, E., Lang, B., von Lützow, M., Marin-Spiotta, E., van Wesemael, B., Rabot, E., Ließ, M., Garcia-Franco, N., Wollschläger, U., Vogel, H. and Kögel-Knabner, I. 2019. Soil organic carbon storage as a key function of soils-A review of drivers and indicators at various scales. *Geoderma* 333, 149–162.

Williams, C. A., Gu, H., MacLean, R., Masek, J. G. and Collatz, G. J. 2016. Disturbance and the carbon balance of US forests: a quantitative review of impacts from harvests, fires, insects, and droughts. *Global and Planetary Change* 143, 66–80. https://doi.org/10.1016/j.gloplacha.2016.06.002.

Zabowski, D., Chambreau, D., Rotramel, N. and Thies, W. G. 2008. Long-term effects of stump removal to control root rot on forest soil bulk density, soil carbon and nitrogen content. *Forest Ecology and Management* 255(3–4), 720–727.

Zak, D. R., Pellitier, P. T., Argiroff, W., Castillo, B., James, T. Y., Nave, L. E., Averill, C., Beidler, K. V., Bhatnagar, J., Blesh, J., Classen, A. T., Craig, M., Fernandez, C. W., Gundersen, P., Johansen, R., Koide, R. T., Lilleskov, E. A., Lindahl, B. D., Nadelhoffer, K. J., Phillips,

R. P. and Tunlid, A. 2019. Exploring the role of ectomycorrhizal fungi in soil carbon dynamics. *New Phytologist* 223(1), 33–39.

Zhang, B., Zhou, X., Zhou, L. and Ju, R. 2015. A global synthesis of below-ground carbon responses to biotic disturbance: a meta-analysis. *Global Ecology and Biogeography* 24(2), 126–138. https://doi.org/10.1111/geb.12235.

Zhang, X., Guan, D., Li, W., Sun, D., Jin, C., Yuan, F., Wang, A. and Wu, J. 2018. The effects of forest thinning on soil carbon stocks and dynamics: a meta-analysis. *Forest Ecology and Management* 429, 36–43.

Zimmermann, S. and Frey, B. 2002. Soil respiration and microbial properties in an acid forest soil: effects of wood ash. *Soil Biology and Biochemistry* 34(11), 1727–1737. https://doi.org/10.1016/S0038-0717(02)00160-8.

Chapter 2

The contribution of agroforestry systems to improving soil carbon sequestration

Lydie-Stella Koutika, Research Centre on the Durability and the Productivity of Industrial Plantations (CRDPI), Republic of the Congo; Nicolas Marron, UMR 1434 Silva, INRAE Grand-Est Nancy, Université de Lorraine, AgroParisTech 54000 Nancy, France; and Rémi Cardinael, AIDA, University of Montpellier, CIRAD, Montpellier, France, CIRAD, UPR AIDA, Harare and University of Zimbabwe, Zimbabwe

1 Introduction

Considered as a climate-smart production system by the FAO (FAO, 2010), agroforestry systems (AFSs) are defined as complex agro-ecosystems combining trees and crops, or trees and pastures, within the same field. Agroforestry is a generic term that includes a wide variety of systems differing in tree and crop species arrangements depending on climate zone and region, making AFS classification challenging (Somarriba, 1992; Torquebiau, 2000). AFSs in the tropics often include livestock (FAO, 2010). AFSs are sometimes defined as 'trees outside forests'. Under this definition, AFSs may be present on about 46% of all agricultural lands supporting 30% of global rural populations (FAO, 2010; Zomer et al., 2016). AFSs cover approximately 1 billion hectares of land and contribute to the livelihoods of over 900 million people (Zomer et al., 2014). A review of a total of 72 scientific, peer-reviewed articles linked to biomass carbon (C) and soil organic C

http://dx.doi.org/10.19103/AS.2022.0106.19

(SOC) storage identified eight main types of AFS comprising alley cropping fallows, hedgerows, multistrata systems, parklands, shaded perennial crops, and silvoarable and silvopastoral systems (Cardinael et al., 2018a), but also much more complex classifications exist (Nair, 1985). Cardinael et al. (2018a) defined alley cropping as intercropping with small shrubs, often N_2-fixing trees, which are regularly pruned or coppiced, to improve soil fertility; this system is mainly found in tropical areas. In contrast, silvoarable systems are defined as arable crops associated with mainly large trees, used either for fuel/firewood production (poplar, for instance) or for timber wood (walnut trees, for instance); these systems are often found in temperate ecosystems, but also exist in subtropical India and China. In temperate regions, hedgerows, windbreaks and silvopastoral systems are the most widespread AFS. In the dry tropics, parklands (e.g. *Faidherbia albida* (Delile) A. Chev.) and fallows are common, while in humid tropics, complex agroforests and shaded perennial-crop systems are widespread.

Farming using AFS has been accepted as a C sequestration activity under the afforestation and reforestation approach (Albrecht and Kandji, 2003; Nair et al., 2009a,b) since the United Nations Framework Convention on Climate Change (UNFCCC) includes the use of C sequestration through afforestation and reforestation in greenhouse-gas offsetting (http://unfccc.int/essential _background/glossary/items/). AFSs have also been included as a greenhouse gas (GHG) mitigation strategy under the Kyoto Protocol since 2007 (Stavi and Lal, 2013; Baah-Acheamfour et al., 2017). The Kyoto Protocol was adopted on 11 December 1997 and entered into force on 16 February 2005, and has 192 signatory Parties (https://unfccc.int/kyoto_protocol). The Protocol initiated by the UNFCCC was designed to accompany industrialized countries, recognized as largely responsible for the current high levels of GHG emissions in the atmosphere, in their transition to regulate and lower GHG emissions in accordance with agreed individual targets.

Indeed, farming using AFSs is a promising C sequestration strategy according to the Intergovernmental Panel for Climate Change (IPCC, 2000, 2019). The C sequestration potential of AFS was estimated to be around 600×10^6 Mg C year^{-1} on an area of 630 million ha of infertile croplands by 2040, while the potential of other nature based solutions was much lower (i.e. 20×10^6 Mg C year^1 for wetland restoration, 50 Mg C year^1 for restoration of degraded land, 250 Mg C year^1 for forest management, 375 Mg C year^1 for grazing management, 20 Mg C year^1 for rice management and 150 Mg C year^1 for cropland management (IPCC, 2000; Jose, 2009; Abbas et al., 2017; Chapman et al., 2020). Between 2000 and 2010, tree cover on agricultural lands increased by 3.7% worldwide, resulting in an increase of more than 2×10^9 Mg C (2 Pg C) of stored biomass carbon (Zomer et al., 2016). AFS can also contribute to reducing the pressure on natural forests (Asaah et al., 2011; Getnet, 2020).

Direct and indirect pathways for C sequestration in soils can be distinguished (Nair et al., 2009b). The conversion of CO_2 into soil inorganic C compounds (calcium and magnesium carbonates) during inorganic chemical reactions is a direct process (see Chapters 7 and 23 of this book). Organic C accumulation by plants during photosynthesis and the subsequent transfer of C to the soil during the decomposition of plant material is an indirect pathway (Nair et al., 2009b; Chapter 3 of this book). AFSs improve SOC sequestration (Feliciano et al., 2018; Cardinael et al., 2018a) through indirect pathways, and also enhance soil faunal and microbial diversity, abundance and activity (Sileshi and Mafongoya, 2006; Marsden et al., 2019; Cardinael et al., 2019). When N_2-fixing species (NFS) are used in AFSs, SOC sequestration may be boosted by enhanced microbial activity in the litter and soil, thus favouring litter mineralisation (Bini et al., 2012; Santos et al., 2017a,b; Ortaş et al., 2017; Wang et al., 2017) and improving C cycling and SOC sequestration efficiency (Bini et al., 2012, 2013; Pereira et al., 2017, 2018). AFS have great potential to (1) increase SOC accumulation and ease adaptation and resilience to climate change (IPCC, 2000; Albrecht and Kandji, 2003; Verchot et al., 2007; Nair et al., 2009a; Nguyen et al., 2013; Mbow et al., 2014; van Noordwijk et al., 2021; Cardinael et al., 2021); (2) provide more food and income diversification for the growing populations, mainly in developing countries such as sub-Saharan Africa or Asia (Sileshi et al., 2008; Mbow et al., 2014; Prasad et al., 2016); (3) restore degraded lands (Sileshi et al., 2007; Verchot et al., 2007); and (4) conserve and boost ecosystem biodiversity while simultaneously improving human welfare (Sileshi et al., 2007, 2008; Asaah et al., 2011; Chowdhury et al., 2020).

Reviewing data from 86 published and peer-reviewed studies, Feliciano et al. (2018) argued that time since conversion to AFS was the major factor affecting biomass C sequestration; SOC sequestration, on the other hand, was mostly influenced by climate. Ma et al. (2020) found that previous land use and tree age were the most important drivers of SOC sequestration in AFSs. In a review of 15 publications, Kim et al. showed that the SOC sequestration rate was divided by four by the end of the first year after AFS establishment and then gradually diminished with time while SOC stocks reach a new equilibrium (Kim et al., 2016). Furthermore, AFSs have many advantages linked their ability to improve SOC sequestration, for example, enhancing the overall productivity and efficiency of a land-use system (Nair et al., 2009a).

The '4 per 1000' Initiative encourages the implementation of SOC sequestration practices considering the specificities of the targeted region, such as climate, social and economic factors, but also the barriers, risks and trade-offs (Rumpel et al., 2019). In this context, it should be considered that AFSs are diverse and their SOC sequestration potential is controlled by several factors like topography, climate, geographical area and socioeconomic context (Nair et al., 2009a; Soto-Pinto et al., 2010; Stainback et al., 2012; Feliciano et al., 2018; Corbeels et al., 2019). SOC sequestration in AFSs also strongly depends on species composition and soil characteristics (Asaah et al., 2011; Abbas et al., 2017; Ma et al., 2020).

This chapter shows to which extent climate, geographical regions, soil characteristics, type of AFS and technical factors such as the choice of planted species, tree density and tree management control SOC sequestration, both in terms of the amount of SOC stored and in terms of the SOC form. The chapter also compares SOC sequestration in AFSs and other agroecosystems in different geographical and climatic areas. The main objectives are (1) to highlight how AFSs may improve SOC sequestration relative to other practices or approaches; and (2) to disentangle the main factors determining SOC sequestration processes in AFSs, (3) to identify barriers and (4) provide recommendations on how to improve SOC sequestration in AFS.

2 Improved soil carbon sequestration in agroforestry relative to other systems

SOC sequestration potential in AFSs may be evaluated in comparison to other practices such as afforestation, reforestation or agriculture without trees (Nair et al., 2009a). Studies have shown that AFSs may sequester C in both soil and biomass and that they improve soil quality more than agricultural systems without trees (Feliciano et al., 2018; Chowdhury et al., 2020; Gusli et al., 2020; Ma et al., 2020). Abbas et al. (2017) argued that AFS maintain greater amounts of C in both above- and belowground biomass compared to grazing and pure cropping systems. AFSs do indeed boost SOC sequestration while improving soil quality (Isaac et al., 2012; Feliciano et al., 2018). Chowdhury et al. (2020) reported higher concentrations of soil organic matter (SOM), available phosphorus (P) and exchangeable potassium (K) in AFSs (4.75%, 12.17 µg g¹, 0.39 mg kg¹, respectively) than after reforestation (3.18%, 6.50 µg g¹ and 0.21 mg kg¹, respectively) or in slash-and-burn plots (1.83%, 5.90 µg g¹ and 0.03 mg kg¹, respectively) in the hilly regions of south-eastern Bangladesh. Globally, SOC stocks increase following land-use changes from less complex systems such as pure cropping systems to more complex ones like AFSs. The greater increase in SOC stocks found in AFSs converted from previous croplands compared to AFSs converted from previous grasslands may be due to the lower initial SOC stock found in croplands. Increasing SOC stocks is a slow process; it usually takes several years, or even decades, after a land-use change before a measurable change in SOC stocks can be observed (Poeplau et al., 2011; Ma et al., 2020).

Several recent meta-analyses have quantified the carbon sequestration potential of the AFS as compared to other land-use systems. de Stefano and Jacobson (2018) compared the C sequestration potential of transitioning from diverse land uses toward an AFS. A decrease in SOC stocks was reported for a conversion from forest to AFS (−26% and 24% at 0-15 cm and 0-30 cm soil depths, respectively), while SOC stocks after conversion from agriculture to AFS increased (+26%, 40% and 34% at 0-15 cm, 0-30 cm and 0-100 cm soil depths, respectively). Even though several meta-analyses have shown no difference in SOC for grasslands converted to silvopasture or even a loss of SOC in temperate regions (Mayer et al., 2022), an increase of 9% and 10% at 0-10 cm and 0-30 cm soil depths has been reported globally after grassland/pasture conversion to AFS, but also after conversion from uncultivated or other land uses to AFS (+25% at 0-30 cm with, however, a decrease of −23% at the 0-60 cm soil depth) (de Stefano and Jacobson, 2018). Indeed, silvopastoral AFSs generally do have the highest SOC stocks. Another global meta-analysis reported the highest increase in SOC stocks for silvopastoral AFSs compared to croplands or pastures (Shi et al., 2018). However, exceptions exist; for instance, under lowland humid tropical and subtropical conditions, multistrata AFSs may have the potential to sequester 12.6% more SOC than forests (Chatterjee et al., 2018, Fig. 1).

AFSs can store large amounts of C in both above- and belowground compartments (Cardinael et al., 2018a). Nevertheless, it is a well-documented fact that natural forests have a greater potential to sequester C than AFS. Indeed, they lose SOC and biomass C when converted to an AFS (Cardinael et al., 2018a; Chatterjee et al., 2018; Feliciano et al., 2018; de Stefano and Jacobson, 2018). The following land-use ranking in terms of average SOC stocks can be found in the literature: old forests (288 Mg C ha^{-1}) > mixed-species reforestation (181 Mg C ha^{-1}) > tree plantations (142 Mg C ha^{-1}) > grasslands (110 Mg C ha^{-1}) > AFS (99 Mg C ha^{-1}) > cropland (40 Mg C ha^{-1}) (Pandey, 2002; Albrecht and Kandji, 2003). Based on data reported in the literature for all climates and AFS types combined, Shi et al. (2018) also calculated a mean SOC stock in the AFS (at a 1-m soil depth) of 126 Mg C ha^{-1}, which is 19% more than in treeless agricultural systems. Comparing SOC sequestration among systems has also shown that sequestration potential depends on geographical zone. Chatterjee et al. (2018) showed that, compared to treeless agricultural systems, SOC stocks under AFS were higher by 27% in arid and semi-arid regions, by 26% in lowland humid tropics and by 5.8% in Mediterranean regions, and that they were lower by 5.3% in temperate conditions in the 0-100 cm soil layer (Fig. 1).

Tree root systems may induce C storage in very deep soil layers (Nair et al., 2009a; Ma et al., 2020; Cardinael et al. (2015b, 2018b). This deep rooting of agroforestry trees may be due to the competition with associated crops and to soil tillage (Cardinael et al., 2015a). SOC in AFSs may be sequestered in

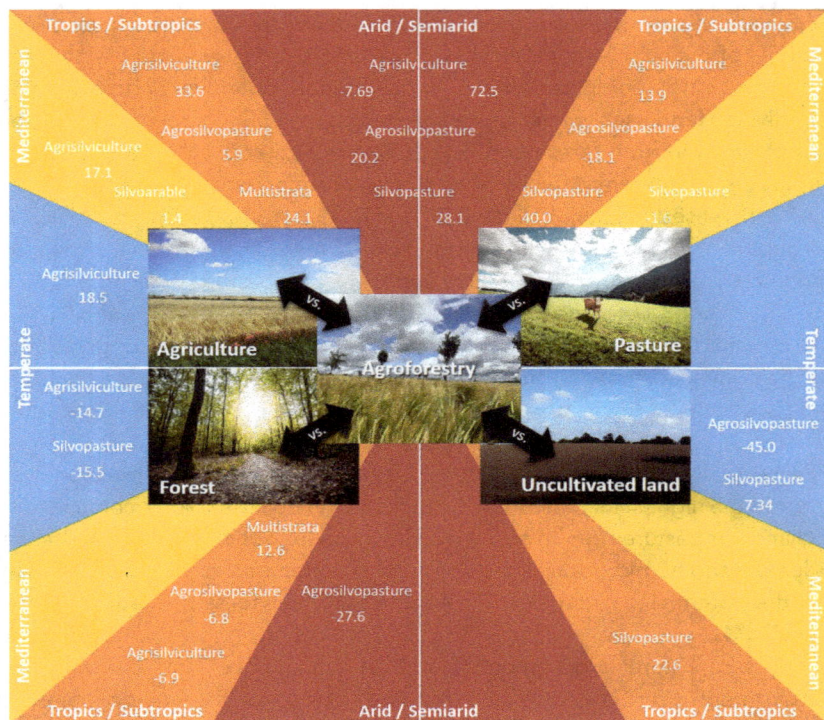

Figure 1 Percentages of variation in SOC stocks (to 40 cm in depth) when AFSs are compared to agriculture, pasture, forest or uncultivated land in different agro-ecological regions, according to the meta-analysis by Chatterjee et al. (2018). Positive values indicate higher SOC stocks under AFS.

deeper layers since tree root systems can grow very deep, and this, in turn, results in a higher input of organic matter to the deeper soil layers (Nair et al., 2010; Germon et al., 2016). At the same time, high surface input of organic matter (e.g. leaf litter) from trees favours the production of dissolved organic C that can be transported to deeper soil layers, thereby contributing to subsoil C storage (Lorenz and Lal, 2005; Ma et al., 2020). In their meta-analysis, Shi et al. (2018) showed that SOC stocks were higher in AFSs (averaged over all AFS types) than in croplands in all soil layers, though the difference was only significant for the 60–80 soil layer. Nygren et al. (2012) argued that root nodule dynamics and N rhizo-deposition from N_2-fixing species in AFSs are sources of readily available N to crops and indirectly affect the dynamics of SOC sequestration (Van Groningen et al., 2017). However, the authors reported that these processes have been poorly investigated to date, even though they highlight the interactions between tree N rhizo-deposition to soil via exudates and root turnover.

Other soil factors play a role in the more efficient SCS in AFSs compared to more conventional agricultural systems. Several studies have shown an increase in soil aggregate stability in AFSs (Udawatta et al., 2008), and this should allow better stabilization and protection of soil C. Moreover, the abundance, biomass and diversity of earthworms are also higher in AFSs than in agricultural plots (Hauser, 1993; Price and Gordon, 1998; Cardinael et al., 2019), and this macrofauna plays a crucial role in nutrient cycling, litter decomposition and SOC stabilization in AFSs (Sileshi and Mafongoya, 2006; Marsden et al., 2019).

3 Factors driving soil carbon sequestration in agroforestry systems

SOC sequestration in AFSs is linked to various factors (Nair et al., 2009b). Four main groups of strongly interconnected factors can be identified: (1) climatic and topographic conditions (altitude, climate); (2) plant characteristics (tree species, diversity and age of planted material) and stand characteristics (tree density); (3) management factors (tillage, fertilization, residue management, pruning, thinning, harvesting regimes and holding size); and (4) soil characteristics (texture, structure, nutrient status, and physical, chemical and biological conditions). In addition, the effect of an AFS on SOC stocks increases with tree age and development (Ma et al., 2020).

3.1 Climatic and topographic conditions

The impact of climate is decisive for the SOC sequestering efficiency of an AFS. Some meta-analyses have reported more effective SOC sequestration in AFSs in the tropics than in temperate areas (Feliciano et al., 2018; Shi et al., 2018; Ma et al., 2020). Higher SOC sequestration has also been reported in temperate and tropical zones than in boreal or arid regions (Nair et al., 2009b; Ma et al., 2020), while SOC sequestration in the subtropical regions tends to be greater than in their tropical counterparts (Shi et al., 2018). The potential of sequestering C aboveground is also estimated to be greater in the tropics than in temperate AFSs (Abbas et al., 2017; Feliciano et al., 2018), while the contrary is observed for SOC sequestration (Fig. 2). Aboveground C sequestration in both temperate and tropical zones is higher than in boreal and arid areas (Ma et al., 2020). It takes about 2.5 times longer to sequester C in arid and semi-arid regions (intercropped tree systems) than in humid tropical regions (multistrata shaded perennial and silvopastoral systems), with the latter sequestering up to 20 times more C than the former (Nair et al., 2009b). Moreover, differences in SOC sequestration potential within agro-ecological regions in the tropics have been found (Nair et al., 2009b). SOC sequestration potential (down to 1 m in soil depth) after about 25 years ranges from 5 Mg C ha^1 to 10 Mg C ha^1 (intercropped systems with trees) in

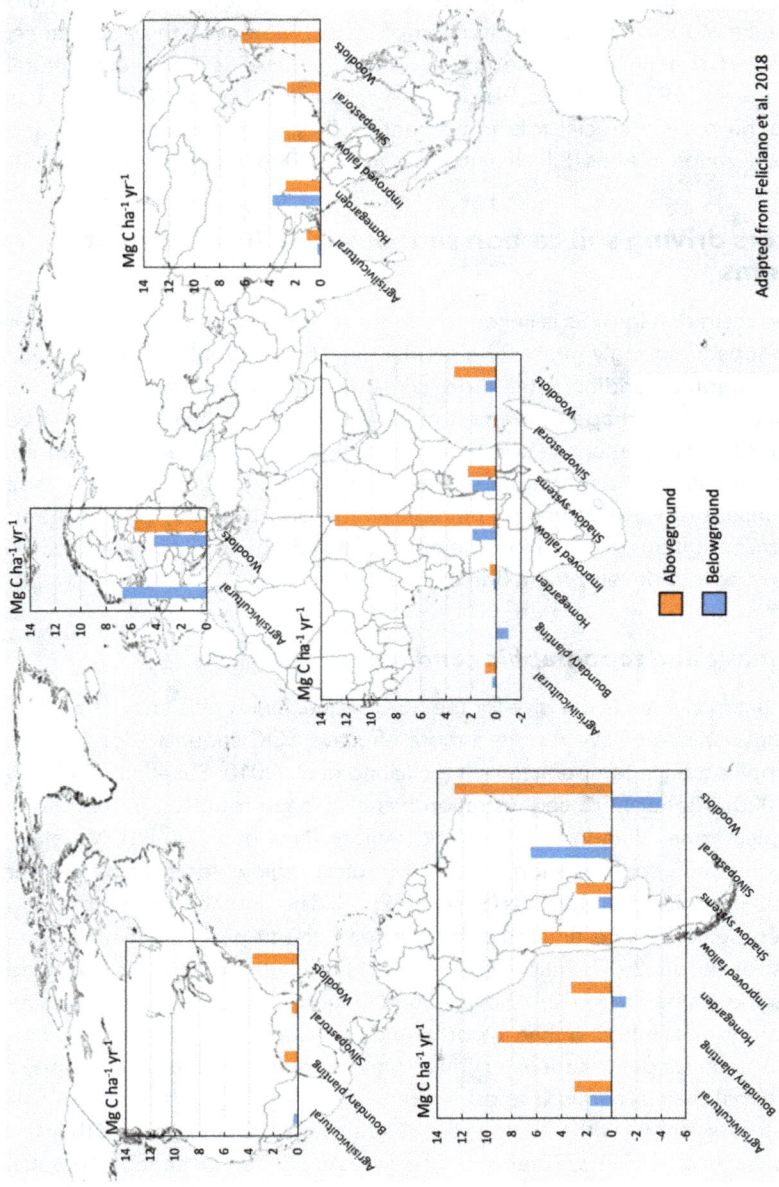

Figure 2 Mean above- and belowground biomass C sequestration (Mg C ha^{-1} year^{-1}) by AFS type and world region according to the meta-analysis by Feliciano et al. (2018).

the arid and semi-arid lands while in humid tropics, the potential ranges from 100 Mg C ha[1] to 250 Mg C ha[1] (multistrata shaded perennial systems and home gardens) after about 10 years (Nair et al., 2009b). In their meta-analysis, Feliciano et al. (2018) showed that the benefits of the AFS are greater in tropical climates than in other climates, both in terms of SOC and aboveground C stocks. AFSs in tropical zones have the ability to increase SOC quickly, whereas in temperate zones, SOC usually increases at a slower rate, though it peaks at a higher overall level than in other climates (Ma et al., 2020). Primary productivity, plant biomass and soil C input are typically greater in wet than in dry climates, whereas higher temperatures are associated with greater rates of decomposition and reduced SOC storage. Globally, C accumulation in both the biomass and the soil components in AFSs tends to differ among climatic zones, with overall positive effects from a high mean annual precipitation (MAP) and negative effects from a high mean annual temperature (MAT) (Ma et al., 2020).

3.2 Plant and stand characteristics

In conjunction with pedoclimatic factors, the ability of an AFS to store SOC also depends on the type of agroforestry implemented (choice of tree species, species diversity, etc. – Cardinael et al., 2018a; Feliciano et al., 2018; Hübner et al., 2021), although there may often be an interaction between agroforestry type and region or climate. Eight main AFSs were identified in a systematic review on above- and belowground C storage and SOC in AFSs (Table 1, Cardinael et al., 2018a).

Table 1 Mean aboveground (ABG) and belowground (BLG) biomass carbon accumulation rates (Mg C ha[1] year[1], except for hedgerows: Mg C km[1] year[1]), with standard deviations, for all regions combined

	Number of studies	Tree density per hectare	ABG C accumulation rate	BLG C accumulation rate
Alley cropping systems	90	8568 ± 8403	2.37 ± 1.45	0.55 ± 0.34
Fallows	69	6074 ± 4529	4.42 ± 2.86	1.49 ± 1.56
Hedgerows	15	979 ± 824	0.79 ± 0.69	0.20 ± 0.18
Multistrata systems	51	929 ± 901	3.25 ± 2.54	0.80 ± 0.60
Parklands	7	152 ± 102	0.59 ± 0.46	0.21 ± 0.11
Shaded perennial-crop system	28	4236 ± 2347	2.40 ± 1.54	0.55 ± 0.36
Silvoarable systems	36	574 ± 440	2.56 ± 2.66	0.63 ± 0.67
Silvopastoral systems	28	1329 ± 981	2.70 ± 3.33	0.76 ± 0.89

Source: Adapted from Cardinael et al. (2018a).

(1) Alley cropping systems (or intercropping systems) are usually found in tropical regions and comprise fast-growing, usually leguminous, woody species (mainly shrubs) grown in crop fields, usually at high densities. The woody species are regularly pruned and the pruning residues are applied as mulch on the alleys as a source of organic matter and nutrients. Multistrata systems are multistorey combinations of a large number of various trees at high densities and perennial and annual crops. They include (2) home gardens and (3) agroforests and are mainly found in the humid tropics. In these regions, shaded perennial-crop systems are also common for shade-tolerant species such as cacao and coffee, grown under or between overstory shade trees that can be used for timber or other commercial products. In dry tropics like the Sahel, (4) parklands are widespread. In these systems, agricultural crops or grazing areas occur under scattered mature trees (e.g. *Faidherbia albida*). (5) Fallows are a type of sequential AFS including both natural and improved fallows.

In temperate regions, (6) hedgerows and (7) silvopastures are the most common AFSs. Hedgerows consist of linear plantations around the fields and also include shelterbelts, windbreaks and live fences. Silvopastures are systems where woody species are planted on permanent, often grazed, grasslands. Hedgerows and silvopastures are also commonly found in tropical regions. (8) Silvoarable systems comprise woody species planted in parallel rows at low densities, alternating with an annual crop. This system allows the use of mechanised equipment and, the trees can be harvested for timber (e.g. *Juglans* spp.), but also for fuel (e.g. *Populus* spp.). According to the meta-analysis by Feliciano et al. (2018), silvopastoral AFSs have high SOC stocks whereas in improved fallows, C sequestration in the aboveground biomass was preponderant. The SOC sequestration gained by planting trees in pastures is almost null due to high initial SOC stocks, some authors even reported a SOC loss following conversion of grasslands to silvopastures in temperate regions (Mayer et al., 2022).

In their review of the literature, Pellerin et al. (2020) calculated a mean SOC storage rate of 0.25 Mg C ha^1 year^1 for silvoarable systems under temperate conditions in North American and Europe (between −0.23 Mg C ha^1 year^1 and +0.73 Mg C ha^1 year^1, for a mean soil depth of 36.4 cm, a mean tree density of 182 trees per ha and a mean tree age of 15.8 years; 25 study cases in 11 publications). This value is close to the one found by Cardinael et al. (2017) for five silvoarable systems in France (a mean of 0.24 Mg C ha^1 year^1, with a range of 0.09 Mg C ha^1 year^1 to 0.46 Mg C ha^1 year^1, for the 0–30 cm soil layer, a mean tree density of 75 trees par ha, and a mean tree age of 17.8 years). However, significant additional SOC sequestration in silvoarable systems has also been observed in the subsoil down to 1 m in depth (Cardinael et al., 2015b; Shi et al., 2018). For silvopastoral systems located in Europe and North America, Pellerin et al. (2020) found no significant effect of tree plantation on SOC stocks in perennial grasslands, with a mean value of 0.05 Mg C ha^1 year^1 (between −0.20 Mg C ha^1 year^1 and 0.29 Mg C ha^1 year^1, for a mean soil depth of 42.9 cm, a mean tree density of 272 trees per ha and a mean tree age of 25.3 years; 12 study cases in 7 publications). Finally, for hedgerows, the mean SOC storage rate was 0.75 Mg C ha^1 hedgerow year^1 (between 0.49 Mg C ha^1 hedgerow year^1 and 1.02 Mg C ha^1 hedgerow year^1, for a mean soil depth of 38.8 cm, a mean tree density of 739 trees per km and a mean tree age of 26.6 years). The same ranking was found for temperate AFS by Mayer et al. (2022) in their meta-analysis: hedgerows (0.32 ± 0.26 and 0.28 ± 0.15 Mg C ha^1 year^1 for topsoils and subsoils, respectively) > alley cropping systems (0.26 ± 1.15 and 0.23 ± 0.25 Mg C ha^1 year^1) > silvopastoral systems (−0.17 ± 0.50 and 0.03 ± 0.26 Mg C ha^1 year^1).

Tree diversity also has an impact on C sequestration. AFSs with multiple tree species contain greater biomass C stocks and accumulate biomass C faster than systems with only one tree species. Including multiple tree species in an AFS may promote a more efficient use of resources compared with a single-species AFS, thereby leading to greater net primary production and, consequently, greater C sequestration in above- and belowground biomass (Ma and Chen, 2017). Given that the positive effect of tree species diversity on biomass production increases over time as a result of increasing interspecific complementarity (Reich et al., 2012), the general pattern of biomass C accumulation as a function of tree age is expected to differ between AFSs with multiple tree species and those with a single tree species (Ma et al., 2020). Under temperate conditions, it has also been shown that SOC sequestration tended to be higher for AFS with broadleaf tree species compared to coniferous species (Mayer et al., 2022).

3.3 Management factors

Trees generally grow faster in AFSs than in forests due to weaker intra-specific competition and to fertilising inputs (Balandier and Dupraz, 1998). Thus, each individual tree's contribution to SOC storage is much greater (Pellerin et al., 2020). In the AFS, agricultural inputs concern fertilisers (chemical or organic amendments: manure, wastewater, sewage sludge, etc.) and/or animal dung and urine which also stimulate tree growth and their subsequent ability to store C. Fertilisation can also occur through symbiotic N_2 fixation, particularly when either herbaceous or woody species of the *Fabaceae* family (legumes) are a component of the AFS.

Introducing N_2-fixing species (NFS) in an AFS can lead to a more N self-sufficient system, as argued by Prasad et al. (2016) in a review on AFSs linked to food production and security in India, a country with a very long history of agroforestry practices. The importance of N in the SOC sequestration process is well known; it results from the interaction among the C, N and P cycles as well as constraints linked to stoichiometry (Van Groningen et al., 2017). In one study, the total amount of N_2 biologically fixed from the atmosphere by the inoculated *Leucaena leucocephala* (Lam) de Wit was estimated to be more than 500 kg of N ha[1] year[1] for the subsequent maize crop (Sanginga et al., 1986). Almost the same range was reported in another study: up to 472 kg N ha[1] year[1] for *L. leucocephala*, *Gliricidia sepium* and *Calliandra calothyrsus*, though lower values were found for *Acacia melanoxylon* and *Acacia holoserica* (<50 kg N ha[1] year[1]) (Giller, 2001).

Integrating NFS in AFSs not only improves soil N status, but also enhances the AFS's potential to sequester C and improve soil health and quality (Nair et al., 2009a; Kaonga, 2005; Lorenz and Lal, 2014; Sebukyu and Mosango, 2012; Dubiez et al., 2019). Enhanced SOC sequestration in NFS-based agroforestry may be attributed to the decomposition of large quantities of high-quality organic matter from N_2-fixing trees (von Haden et al., 2019). NFS may also transfer the atmospheric N_2 they fix to non-NFS species (Epron et al., 2013; Paula et al., 2015). This was demonstrated for pot culture on red Mediterranean soils where legume tree species facilitated N transfer to interspersed crops stimulated by high P conditions, thus evidencing the effective nutrient economy of a properly managed NFS-based AFS (Isaac et al., 2012). Under field conditions in north-eastern France, a succession of alfalfa and clover crops grown together with poplars significantly enhanced soil N, thus increasing growth and growing season length for the trees, compared to poplars in a monoculture (Thomas et al., 2020).

In most rural populations, the concept of increasing SOC sequestration to improve soil health and quality is unknown; most farmers verify and evaluate the benefits of practices through crop yields and income (ASB, 2000; Schroth and Sinclair, 2003; Degrande et al., 2007). To stimulate interest in adopting new methodologies, a project in north and northwest Cameroon promoted an agroforestry approach with N_2-fixing trees and selected cultivars of local fruit and nut trees with local and regional markets as using 'fertilizer trees' to improve soil quality and obtain higher incomes (Asaah et al., 2011).

3.4 Soil characteristics

The success of implementing an AFS obviously depends on the soil characteristics of the target area (Nair et al., 2009a,b; Hübner et al., 2021). The benefits of an AFS may be more perceptible in unproductive or degraded lands than on richer or rehabilitated parcels, especially when N_2-fixing trees are used (Bisiaux et al., 2009; Chowdhury et al., 2020). An agroforestry project of 8000 hectares including *Acacia auriculiformis* A. Cunn. ex Benth. established from 1987 to 1992 on the infertile sandy soils of the Batéké plateau in the Democratic Republic of Congo led to both improved soil quality (increased C and N concentrations and increased soil acidity) and an increase in crop yields (cassava, maize) and honey production (Bisiaux et al., 2009; Nsombo, 2016). The higher SOM turnover than commonly observed in similar ecosystems were linked to soil intrinsic characteristics in addition to the higher quality and quantity of aboveground biomass production. There may also be higher SOM decomposition or mineralisation in tropical ecosystems due to high temperatures and mean annual rainfall (Hassink, 1997; Koutika et al., 1999). These processes may be enhanced in ecosystems with fragile soils with a limited ability to store and protect added SOM due, for instance, to their coarse texture or low nutrient availability (Derrien et al., 2014). The ability of the silt and clay particles in the soil to protect organic material, obviously depends on intrinsic soil characteristics (Hassink, 1997; Koutika et al., 1999) and is strongly linked to physical soil characteristics (Nair et al., 2009a,b; Chowdhury et al., 2020). Coarse textured and/or disturbed soils contain few organic materials and have a poor structure and aggregation (Elliott and Coleman, 1988; Six et al., 2000). In addition to the above-mentioned benefits on soil attributes observed in AFSs, especially when NFS are introduced, AFS have a positive impact on the diversity and functioning of soil invertebrates (Sileshi and Mafongoya, 2006; Marsden et al., 2019). The combined soil benefits provided by implementing an AFS usually lead to enhanced faunal and microbial activity, more robust arbuscular mycorrhizal fungi and better nutrient cycling (Bini et al., 2012, 2013; Pereira et al., 2017, 2018, 2020; Ortaş et al., 2017; Wang et al., 2017; Dubiez et al., 2019).

4 Other co-benefits of sequestering soil carbon in agroforestry systems

Food availability and security are two of the expected co-benefits of improving SOC sequestration through AFSs, which could benefit poorer populations in some regions of Africa or Asia (Mbow et al., 2014; Prasad et al., 2016). For certain types of AFSs, increases in crop yields can be expected. Food security occurs through increased crop yields following an improvement in soil quality (Sileshi et al., 2007, 2008). Yields for maize, a staple crop in many African countries, increased from 1.3 Mg ha[1] year[1] to 1.6 Mg ha[1] year[1] in AFSs in sub-Saharan Africa as compared to pure cropping systems (Sileshi et al., 2008). In the case mentioned above (Sileshi et al., 2008), improved fallows and coppicing woody legumes were key as there was a temporal association with no competition. This also occurs in parklands with *Faidherbia albida* where no competition occurs since the trees bear leaves only during the dry season when there are no crops; in addition, soil N status is improved through N_2 fixation by the NFS. However, in most cases, one can expect competition to be higher than facilitation and crop yields to decrease in an AFS. In addition to the above-mentioned benefits of SOC sequestration in terms of improved soil fertility and

Table 2 Improved SOC sequestration in agroforestry systems and co-benefits in different parts of the world according to the literature

Region and climate	Type of paper	Benefits of improved SOC sequestration			Other co-benefits of improved C sequestration			References
		Soil fertility and land restoration	Food availability and security	Adaptation to climate change and resilience	Fuel wood energy	Forest products	C payment markets	
Overall	Review	+	+	+	+			Abbas et al. (2017)
Cameroon	Review	+	+	+	+	+		Asaah et al. (2011)
Tropics	Review	+	+	+	+			Albrecht and Kandji (2003)
Overall	Review	+					+	Cardinael et al. (2018a)
Bangladesh	Research	+		+				Chowdhury et al. (2020)
DR Congo	Research	+	+		+	+		Dubiez et al. (2019)
Overall	Meta-analysis	+	+	+	+	+		Feliciano et al. (2018)
Ethiopia	Review			+			+	Getnet (2020)
Indonesia	Research	-	+	+				Gusli et al. (2020)
Central Africa	Research						-	Lescuyer et al. (2009)
Overall	Review	+	+	+	+	+		Lorenz and Lal (2014)
Africa	Review	+	+	+	+	+		Mbow et al. (2014)
Overall	Review	+	+	+			+	Nair et al. (2009a,b)
India	Review	+	+	+	+	+		Prasad et al. (2016)
Uganda	Research	+	+	+	+	+		Sebukyu and Mosango (2012)
Eastern and Southern Africa	Review	+	+	+	+	+	+	Sileshi et al. (2007)
Mexico	Research	+	+	+		+		Soto-Pinto et al. (2010)
Rwanda	Review	+	+			+	+	Stainback et al. (2012)
Overall	Review			+				Verchot et al. (2007)

biodiversity, AFSs can also provide tree products, i.e. fodder, fuelwood, food and building materials (Sileshi et al., 2007, 2008; Nair et al., 2009a; Lescuyer et al., 2009; Sebukyu and Mosango, 2012; Gusli et al., 2020, Table 2). Other ecosystem services such as land restoration and biodiversity protection are also provided by AFSs (Lescuyer et al., 2009; Sebukyu and Mosango, 2012; Mbow et al., 2014). In some tropical areas, AFSs help to decrease pressure on forests for fuel wood energy (Sileshi et al., 2007; Prasad et al., 2016). C sequestration could also be an economic opportunity for subsistence farmers in developing countries if trading C sequestered through agroforestry activities becomes more widespread (Lescuyer et al., 2009; Nair et al., 2009b; Feliciano et al., 2018).

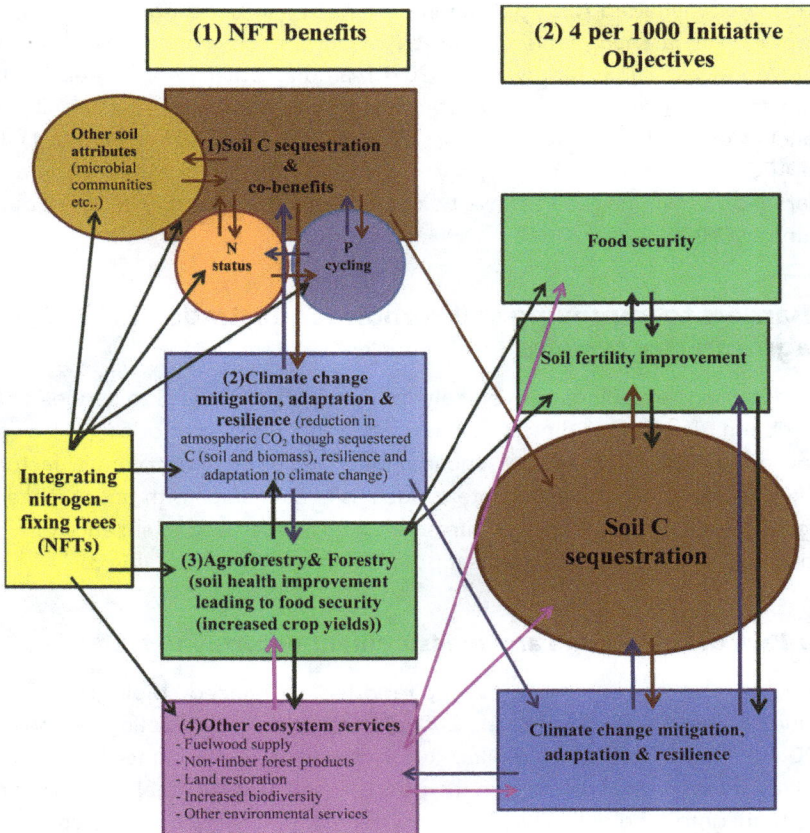

Figure 3 Conceptual diagram showing (1) N-fixing tree (NFT) benefits and links to (2) the '4 per 1000' Initiative objectives: how introducing NFTs in agroforestry and forestry leads to C sequestration and co-benefits in other ecosystem services, and promotes the '4 per 1000' Initiative in the Congo Basin (DR Congo and Rep. Congo). Source: Adapted from Koutika et al. (2021).

The co-benefits of sequestering C in AFSs in Central Africa (mainly in DR Congo and the Republic of the Congo) have been summarized in a review (Koutika et al., 2021). The authors developed a conceptual scheme of the benefits of N_2-fixing trees related to the main objectives of the '4 per 1000' 'Soils for food security and climate' Initiative (Fig. 3). C sequestration, with its links to N and P cycling, microbial communities and other soil attributes, helps to control not only the quantity but also the quality of sequestered SOC (Bini et al., 2012, 2013; Pereira et al., 2018, 2020; Koutika et al., 2019, 2020a,b). Not only climate change mitigation, by reducing atmospheric CO_2 through SOC sequestration, but also improved resilience and adaptation to climate change are also benefits of planting N_2-fixing trees and implementing agroforestry practices (Verchot et al., 2007; Stavi and Lal, 2013; Baah-Acheamfour et al., 2017). Well-managed AFSs play an important role in boosting resilience and adaptation to a changeable climate by creating microclimates, for instance (Nguyen et al., 2013; Mbow et al., 2014), among other benefits related to adaptation and resilience to climate change (Arenas-Corraliza et al., 2018; Blanchet et al., 2021; van Noordwijk et al., 2021; Cardinael et al., 2021). In the AFOLU (Agriculture, Forestry and Other Land Use) sector, mitigation and adaptation to climate change have been identified as a necessity in developing countries (Mbow et al., 2014).

5 Barriers to improving soil carbon sequestration in agroforestry systems

Some barriers lessening, or even halting, the ability of an AFS to sequester C have been identified and may be classified in two main categories: (1) lack of knowledge and measurement uncertainties, and (2) socioeconomic factors (linked to social and political contexts such as relationships with policy makers or governments, lack of funds or infrastructure. This may lead to absence of AFS projects in regions with the greatest potential to sequester C (Table 3).

5.1 Lack of knowledge and measurement uncertainties

Even though studies have often reported enhanced biological SOC sequestration in AFSs (Nair et al., 2009a), estimating or evaluating biological SOC stocks in AFSs may be a difficult task due to limited accuracy (Nair et al., 2009a) or technical constraints in the poorest regions. Cardinael et al. (2018a) also highlighted the difficulty of correctly estimating changes in SOC stocks in AFSs due to the lack of rigorous long-term measurements, as already reported by Nair et al. (2009a) and Nair (2012). Even when SOC is measured in AFSs, it is rarely compared to an adequate reference system (Lorenz and Lal, 2014; Cardinael et al., 2018a; Feliciano et al., 2018).

Table 3 Main barriers to improving SOC sequestration in agroforestry systems

Types of barriers	References
(i) Lack of knowledge, technical barriers Incorrect measurements of sequestered C Lack of ability to measure sequestered C Inappropriate AFS implementation (wrong choice of tree species, animal damage, etc.) Lack of access to plant seedlings, irrigation or soil drainage systems, mechanization, inappropriate field size, etc.	Balandier and Dupraz (1998), Cardinael et al. (2018a), Chatterjee et al. (2018), Feliciano et al. (2018), Lorenz and Lal (2014), Nair et al. (2009a), Shi et al. (2018)
(ii) Socioeconomic barriers Social and political contexts Policymakers (lack of policies) Governments (insufficient support of infrastructure, etc.) Lack of training or extension services Lack of funds, infrastructure or AFS-enabling projects in the regions with the greatest potential to sequester C Land tenure issues	Nair et al. (2009a), Mbow et al. (2014), Stainback et al. (2012), Bucagu et al. (2013), Foundjem-Tita (2013)

Many studies on SOC sequestration in AFSs are based on biomass and soil sampling undertaken at a specific period, others are based on field measurements and others use simulation models based on various assumptions. Almost all the studies to date report the amount of SOC stocks for specific systems and only at the time of study. While these data are extremely valuable for understanding existing conditions, they represent only one point in the continuum of land-use changes. They are not sufficient when the goal is to predict the rate of change in SOC when land-management practices change from agriculture or forest to an AFS. Long-term monitoring information is needed to properly assess the benefits, not only in terms of SOC but also in terms of the economic advantages associated with the adoption of AFS for local farmers (Chatterjee et al., 2018).

Shi et al. (2018) identified the following important knowledge gaps regarding SOC sequestration in AFS: (1) how to quantitatively assess C allocation and stocks by various AFS components, with a special focus on deep soil C and its dependence on tree species and age; (2) how to optimise the relative area allocated to trees and crops within each AFS type to maximise C sequestration, yield, ecosystem services and to improve environmental conditions; and (3) how to develop new remote sensing approaches to distinguish AFSs from the background of forests, plantations and gardens, and to quantify the AFS area. An investigation of several forestry and agroforestry plantations in the Mediterranean area, in dry central temperate plains, in cold wet central uplands and in mild oceanic areas revealed that a wrong choice of species for the prevailing local conditions may cause some AFSs to fail (Balandier and Dupraz, 1998). Finally, a comprehensive understanding of what happens belowground in the AFS is also lacking, for all its dimensions (temporal, spatial, soil depth). Studies integrating AFS functioning as a whole, embracing the very diverse forms of AFS and considering the multifunctionality of these systems are urgently needed (Cardinael et al., 2020).

5.2 Socioeconomic barriers

There are also socioeconomic barriers that must be overcome, mainly in developing or poor countries in sub-Saharan Africa, to enable SOC sequestration improvement (Demenois et al., 2020) through AFS introduction (Nair et al., 2009a; Stainback et al., 2012; Bucagu et al., 2013; Foundjem-Tita et al., 2013). Success in implementing an AFS and its ability to improve soil C sequestration strongly relies on different factors such as land tenure, policies, technical advice and assistance, government support, etc. Stainback et al. (2012) linked AFS success for smallholders in Rwanda to favourable support from the government and the efficient technical capacity to design, implement and monitor carbon projects included in environmental service markets. In order to overcome barriers to adoption, a participatory approach allowing smallholders to share their knowledge and identify key opportunities, problems and constraints seems appropriate (Bucagu et al., 2013). Other barriers can be a lack of funding, infrastructure or specific AFS projects in regions with the greatest potential to sequester C (Mbow et al., 2014; Stainback et al., 2012; Bucagu et al., 2013; Foundjem-Tita et al., 2013). Rosenstock et al. (2019) formulated four recommendations in order to operationalize broad-scale measurement, reporting and verification (MRV) systems for AFSs under the UNFCCC: (1) develop accessible approaches for representation of lands with AFS; (2) create guidelines for AFS reporting to improve relevance to national policy and transparency; (3) develop carbon-stock change and emission databases relevant for reporting requirements; and (4) navigate the institutional arrangements needed to include agroforestry in MNV.

6 Recommendations

To improve the positive impact of AFSs on SOC sequestration, not only the type of AFS but also the environmental, climatic, socioeconomic and political contexts must be considered prior to adoption and implementation of such systems. SOC sequestration is related to soil characteristics (content of other nutrients such as N and P, soil texture, etc.) and strongly linked to the factors mentioned above. The way the AFS is managed will have a crucial impact on the quantities of C stored in the soil. For example, because of the low density of the trees in agroforestry, there is no natural pruning as there is in the forest. Trees must be pruned, not only to produce wood, but also to limit their competition with crops and facilitate the passage of agricultural machinery.

The management of pruning residues can be a lever to increase SOC storage, especially if they are crushed and brought to the ground in the form of ramial chipped wood (Fungo et al., 2021). Moreover, alignments of trees induce, both directly (input of organic matter from the trees) and indirectly (sub-tree herbaceous vegetation), a heterogeneous distribution of soil carbon. The largest SOC stocks are usually observed near trees rather than in the middle of the cultivated inter-row (Bambrick et al., 2010; Cardinael et al., 2015b). Therefore, managing the sub-tree vegetation in silvoarable systems (sowing or spontaneous cover, mowing or not, etc.) seems crucial for SOC storage. Finally, AFSs should also be managed to increase the SOC reservoir by avoiding burning and by minimizing soil disturbance through limited or zero tillage practices and appropriate erosion control measures (Soto-Pinto et al., 2010).

7 Conclusion

In most situations, AFSs store C (especially in deeper soil layers) more efficiently than croplands but less efficiently than forests. Diversified AFSs in terms of tree species are more efficient for SOC sequestration due to the complementarity among species for the use of resources. Regional climate also has a strong impact, with tropical AFSs being, on average, more efficient than temperate AFSs thanks to the higher biomass production and greater soil C sequestration occurring under wetter conditions. SOC sequestration also depends on the type of AFS: in temperate latitudes, literature reviews have shown that silvoarable systems are, on average, more efficient than silvopastoral systems in storing additional C in soils because pastures already have high SOC stocks. The use of woody or herbaceous N_2-fixing species (NFS) can also be a way to enhance SOC sequestration by stimulating the growth of the non-fixing species in the association. Recommendations to improve the probability of successful SOC sequestration in AFSs include (1) carefully matching the selected tree species to local pedoclimatic conditions in order to maximize plant growth and subsequent C storage, and (2) applying good plantation management practices by, for instance, reducing soil disturbance as much as possible, mulching with fragmented tree pruning residues, and adequately managing sub-tree vegetation (the area where the largest SOC stocks are found).

8 Where to look for further information

8.1 Some guidelines for future research

- More studies including socio-economic aspects of agroforestry systems and linked to sustainable development goals of Agenda 2030 of the United Nations.

- Estimate appropriate agroforestry system according to specific agro ecological zone taking into account climate warming.

8.2 Organisations involved with respect to this topic

- CRDPI (Congo), INRAE (France) and CIRAD (France), University of Zimbabwe (Zimbabwe).

9 References

Abbas, F., Hammad, H. M., Fahad, S., Cerdà, A., Rizwan, M., Farhad, W., Ehsan, S. and Bakhat, H. F. (2017). Agroforestry: a sustainable environmental practice for carbon sequestration under the climate change scenarios - a review, *Environ. Sci. Pollut. Res. Int.* 24(12), 11177–11191. DOI: 10.1007/s11356-017-8687-0.

Albrecht, A. and Kandji, S. T. (2003). Carbon sequestration in tropical agroforestry systems, *Agric. Ecosyst. Environ.* 99(1–3), 15–27.

Arenas-Corraliza, M. G., López-Díaz, M. L. and Moreno, G. (2018). Winter cereal production in a Mediterranean silvoarable walnut system in the face of climate change, *Agric. Ecosyst. Environ.* 264, 111–118. DOI: 10.1016/j.agee.2018.05.024.

Asaah, E. K., Tchoundjeu, Z., Leakey, R. R. B., Takousting, B., Njong, J. and Edang, I. (2011). Trees, agroforestry and multifunctional agriculture in Cameroon, *Int. J. Agric. Sustain.* 9(1), 110–119. DOI: 10.3763/ijas.2010.0553.

ASB (2000). Summary report and synthesis of phase II in Cameroon. Alternatives-to-Slash-and-Burn, Nairobi, Kenya, pp. 2–7.

Baah-Acheamfour, M., Chang, S. X., Bork, E. W. and Carlyle, C. N. (2017). The potential of agroforestry to reduce atmospheric greenhouse gases in Canada: insight from pairwise comparisons with traditional agriculture, data gaps and future research, *The Forestry Chronicle* 93(2), 180–189.

Balandier, P. and Dupraz, C. (1998). Growth of widely spaced trees: a case study from young agroforestry plantations in France, *Agroforest. Syst.* 43(1/3), 151–167.

Bambrick, A. D., Whalen, J. K., Bradley, R. L., Cogliastro, A., Gordon, A. M., Olivier, A. and Thevathasan, N. V. (2010). Spatial heterogeneity of soil organic carbon in tree-based intercropping systems in Quebec and Ontario, Canada, *Agroforest. Syst.* 79(3), 343–353.

Bini, D., Figueiredo, A. F., da Silva, M. C. P., de Figueiredo Vasconcellos, R. L. and Cardoso, E. J. B. N. (2012). Microbial biomass and activity in litter during the initial development of pure and mixed plantations of *Eucalyptus grandis* and *Acacia mangium*, *Rev. Bras. Cienc. Solo* 37, 76–85.

Bini, D., Santos, C. Ad, Bouillet, J., Gonçalves, J. LdM. and Cardoso, E. J. B. N. (2013). *Eucalyptus grandis* and *Acacia mangium* in monoculture and intercropped plantations: evolution of soil and litter microbial and chemical attributes during early stages of plant development, *Appl. Soil Ecol.* 63, 57–66.

Bisiaux, F., Peltier, R. and Muliele, J. C. (2009). Plantations industrielles et agroforesterie au service des populations des plateaux Batéké, Mampu, en République Démocratique du Congo, *Bois for trop* 301(3), 21–32.

Blanchet, G., Barkaoui, K., Bradley, M., Dupraz, C. and Gosme, M. (2021). Interactions between drought and shade on the productivity of winter pea grown in a 25-year-old walnut-based alley cropping system, *J. Agron. Crop Sci.*, 1–16. DOI: 10.1111/jac.12488.

Bucagu, C., Vanlauwe, B., Van Wijk, M. T. and Giller, K. E. (2013). Assessing farmers' interest in agroforestry in two contrasting agro-ecological zones of Rwanda, *Agroforest. Syst.* 87(1), 141–158. DOI: 10.1007/s10457-012-9531-7.

Cardinael, R., Mao, Z., Prieto, I., Stokes, A., Dupraz, C., Kim, J. H. and Jourdan, C. (2015a). Competition with winter crops induces deeper rooting of walnut trees in a Mediterranean alley cropping agroforestry system, *Plant Soil* 391(1–2), 219–235. DOI: 10.1007/s11104-015-2422-8.

Cardinael, R., Chevallier, T., Barthès, B. G., Saby, N. P. A., Parent, T., Dupraz, C., Bernoux, M. and Chenu, C. (2015b). Impact of alley cropping agroforestry on stocks, forms and spatial distribution of soil organic carbon - a case study in a Mediterranean context, *Geoderma* 259–260, 288–299. DOI: 10.1016/j.geoderma.2015.06.015.

Cardinael, R., Chevallier, T., Cambou, A., Béral, C., Barthès, B. G., Dupraz, C., Durand, C., Kouakoua, E. and Chenu, C. (2017). Increased soil organic carbon stocks under agroforestry: a survey of six different sites in France, *Agric. Ecosyst. Environ.* 236, 243–255. DOI: 10.1016/j.agee.2016.12.011.

Cardinael, R., Umulisa, V., Toudert, A., Olivier, A., Bockel, L. and Bernoux, M. (2018a). Revisiting IPCC Tier 1 coefficients for soil organic and biomass carbon storage in agroforestry systems, *Environ. Res. Lett.* 13(12), 124020. DOI: 10.1088/1748-9326/aaeb5f.

Cardinael, R., Guenet, B., Chevallier, T., Dupraz, C., Cozzi, T. and Chenu, C. (2018b). High organic inputs explain shallow and deep SOC storage in a long-term agroforestry system - combining experimental and modeling approaches, *Biogeosciences* 15(1), 297–317. DOI: 10.5194/bg-15-297-2018.

Cardinael, R., Hoeffner, K., Chenu, C., Chevallier, T., Béral, C., Dewisme, A. and Cluzeau, D. (2019). Spatial variation of earthworm communities and soil organic carbon in temperate agroforestry, *Biol. Fertil. Soils* 55(2), 171–183. DOI: 10.1007/s00374-018-1332-3.

Cardinael, R., Mao, Z., Chenu, C. and Hinsinger, P. (2020). Belowground functioning of agroforestry systems: recent advances and perspectives, *Plant Soil* 453(1–2), 1–13. DOI: 10.1007/s11104-020-04633-x.

Cardinael, R., Cadis, G., Gosme, M., Oelbermann, M. and van Noordwijk, M. (2021). Climate change mitigation and adaptation in agriculture: why agroforestry should be part of the solution, *Agric. Ecosyst. Environ.* 319, 107555. DOI: 10.1016/j.agee.2021.107555.

Chapman, M., Walker, W. S., Cook-Patton, S. C., Ellis, P. W., Farina, M., Griscom, B. W. and Baccini, A. (2020). Large climate mitigation potential from adding trees to agricultural lands, *Glob. Chang. Biol.* 26(8), 4357–4365. DOI: 10.1111/gcb.15121.

Chatterjee, N., Nair, P. K. R., Chakraborty, S. and Nair, V. D. (2018). Changes in soil carbon stocks across the forest-agroforest-agriculture/pasture continuum in various agroecological regions: a meta-analysis, *Agr. Eco. Environ.* 266, 55–67.

Chowdhury, F. I., Barua, I., Chowdhury, A. I., Resco de Dios, V. and Alam, M. S. (2020). Agroforestry shows higher potential than reforestation for soil restoration after slash-and-burn: a case study from Bangladesh, *Geology Ecology and Landscapes* 6(1), 48–54. DOI: 10.1080/24749508.2020.1743528.

Corbeels, M., Cardinael, R., Naudin, K., Guibert, H. and Torquebiau, E. (2019). The 4 per 1000 goal and soil carbon storage under agroforestry and conservation agriculture systems in sub-Saharan Africa, *Soil Till. Res.* 188, 16–26. DOI: 10.1016/j.still.2018.02.015.

Degrande, A., Asaah, E., Tchoundjeu, Z., Kanmegne, J., Duguma, B. and Franzel, S. (2007). Opportunities for and constraints to adoption of improved fallows: ICRAF's experience in the humid tropics of Cameroon. In: *Advances in Integrated Soil Fertility Management in Sub-Saharan Africa: Challenges and Opportunities* Bationo, A.,Waswa, B., Kihara, J. and Kimetu, J. (Eds), Springer, The Netherlands, pp. 901–909.

Demenois, J., Torquebiau, E., Arnoult, M. H., Eglin, T., Masse, D., Assouma, M. H., Blanfort, V., Chenu, C., Chapuis-Lardy, L., Medoc, J. and Sall, S. N. (2020). Barriers and strategies to boost soil carbon sequestration in agriculture, *Front. Sustain. Food Syst.* 4, 1–14. DOI: 10.3389/fsufs.2020.00037.

Derrien, D., Plain, C., Courty, P. E., Gelhaye, L., Moerdijk-Poortvliet, T. C. W., Thomas, F., Versini, A., Zeller, B., Koutika, L. S., Boschker, H. T. S. and Epron, D. (2014). Does the addition of labile substrate destabilise old soil organic matter?, *Soil Biol. Biochem.* 76, 149–160.

de Stefano, A. and Jacobson, M. G. (2018). Soil carbon sequestration in agroforestry systems: a meta-analysis, *Agroforest. Syst.* 92, 285–299. DOI: 10.1007/s10457-017-0147-9.

Dubiez, E., Freycon, V., Marien, J. M., Peltier, R. and Harmand, J. M. (2019). Long term impact of *Acacia auriculiformis* woodlots growing in rotation with cassava and maize on the carbon and nutrient contents of savannah sandy soils in the humid tropics (Democratic Republic of Congo), *Agroforest. Syst.* 93(3), 1167–1178. 10.1007/s10457-018-0222-x.

Elliott, E. T. and Coleman, D. C. (1988). Let the soil work for us, *Ecol. Bull.* 39, 23–32.

Epron, D., Nouvellon, Y., Mareschal, L., Moreira, R. Me, Koutika, L.-S., Geneste, B., Delgado-Rojas, J. S., Laclau, J. P., Sola, G., Gonçalves, J. LdM. and Bouillet, J. P. (2013). Partitioning of net primary production in *Eucalyptus* and *Acacia* stands and in mixed-species plantations: two case-studies in contrasting tropical environments, *Forest Ecol. Manag.* 301, 102–111.

FAO (Food and Agriculture Organization of the United Nations) (2010). Climate-Smart Agriculture policies, practices and financing for food security, adaptation and mitigation. Available at: http://www.fao.org/docrep/013/i1881e/i1881e00.pdf.

Feliciano, D., Ledo, A., Hillier, J. and Nayak, D. R. (2018). Which agroforestry options give the greatest soil and above ground carbon benefits in different world regions?, *Agric. Ecosyst. Environ.* 254, 117–129. DOI: 10.1016/j.agee.2017.11.032.

Foundjem-Tita, D., Tchoundjeu, Z., Speelman, S., D'Haese, M., Degrande, A., Asaah, E., van Huylenbroeck, G., van Damme, P. andNdoye, O. (2013). Policy and legal frameworks governing trees: incentives or disincentives for smallholder tree planting decisions in Cameroon?, *Small-Scale For.* 12(3), 489–505. DOI: 10.1007/s11842-012-9225-z.

Fungo, B., Wiesmeier, M. and Cardinael, R. (2021). Agroforestry 1: agrisilvicultural systems. In: *Recarbonizing Global Soils: A Technical Manual of Recommended Management Practices (vol. 3), Cropland, Grassland, Integrated Systems and Farming Approaches – Practices Overview*, Food and Agriculture Organization and ITPS, Rome, pp. 474–486. DOI: 10.4060/cb6595en.

Germon, A., Cardinael, R., Prieto, I., Mao, Z., Kim, J., Stokes, A., Dupraz, C., Laclau, J. P. and Jourdan, C. (2016). Unexpected phenology and lifespan of shallow and deep fine roots of walnut trees grown in a silvoarable Mediterranean agroforestry system, *Plant Soil* 401(1–2), 409–426. DOI: 10.1007/s11104-015-2753-5.

Getnet, D. (2020). GHGs reduction capacity of agroforestry systems in tropical Africa: a review, *Curr. Agri. Res. Jour.* 8(3), 166–177. DOI: 10.12944/CARJ.8.3.02.

Giller, K. E. (2001). *Nitrogen Fixation in Tropical Cropping Systems* (2nd edn.), CAB International Publishing, Wallingford.

Gusli, S., Sumeni, S., Sabodin, R., Muqfi, I. H., Nur, M., Hairiah, K., Useng, D. and van Noordwijk, M. (2020). Soil organic matter, mitigation of and adaptation to climate change in cocoa-based agroforestry systems, *Land* 9(9), 323. DOI: 10.3390/land9090323.

Hassink, J. (1997).The capacity of soils to preserve organic C and N by their association with clay and silt particles, *Plant Soil* 191(1), 77–87.

Hauser, S. (1993). Distribution and activity of earthworms and contribution to nutrient recycling in alley cropping, *Biol. Fertil. Soils* 15(1), 16–20.

Hübner, R., Kühnel, A., Lu, J., Dettmann, H., Wan, W. and Wiesmeier, M. (2021). Soil carbon sequestration by agroforestry systems in China: a meta-analysis, *Agric. Ecosyst. Environ.* 315, 107437. DOI: 10.1016/j.agee.2021.107437.

IPCC (2019). Summary for policymakers. In: *Climate Change and Land: an IPCC Special Report on Climate Change, Desertification, Land Degradation, Sustainable Land Management, Food Security, and Greenhouse Gas Fluxes in Terrestrial Ecosystems'* Shukla, P. R., Skea, J., Calvo Buendia, E., Masson-Delmotte, V., Pörtner, H.- O., Roberts, D. C., Zhai, P., Slade, R., Connors, S., van Diemen, R., et al. (Eds). Cite Report – Special Report on Climate Change and Land (ipcc.ch).

IPCC (2000). Land-Use, Land-Use Change and Forestry. Special Report of the Intergovernmental Panel on Climate Change, Cambridge University Press, Cambridge, p. 375.

Isaac, M. E., Hinsinger, P. and Harmand, J. M. (2012). Nitrogen and phosphorus economy of a legume tree-cereal intercropping system under controlled conditions, *Sci. Total Environ.* 434, 71–78. DOI: 10.1016/j.scitotenv.2011.12.071.

Jose, S. (2009). Agroforestry for ecosystem services and environmental benefits: an overview, *Agroforest. Syst.* 76(1), 1–10.

Kaonga, M. L. (2005). Understanding carbon dynamics in agroforestry systems in eastern Zambia. Ph.D. Dissertation, Fitzwilliam College, University of Cambridge, UK, p. 431.

Kim, D. G., Kirschbaum, M. U. F. and Beedy, T. L. (2016). Carbon sequestration and net emissions of CH_4 and N_2O under agroforestry: synthesizing available data and suggestions for future studies', *Agric. Ecosyst. Environ.* 226, 65–78. DOI: 10.1016/j.agee.2016.04.011.

Koutika, L.-S., Chone, T., Andreux, F., Burtin, G. and Cerri, C. C. (1999). Factors influencing carbon decomposition of topsoils from the Brazilian Amazon Basin, *Biol. Fertil. Soils* 28(4), 436–438.

Koutika, L. S., Ngoyi, S., Cafiero, L. and Bevivino, A. (2019). Soil organic matter quality along rotations in acacia and eucalyptus plantations in the Congolese coastal plains, *Forest Ecosyst.* 6, 39. DOI: 10.1186/s40663-019-0197-8.

Koutika, L. S., Cafiero, L., Bevivino, A. and Merino, A. (2020a). Organic matter quality of forest floor as a driver of C and P dynamics in acacia and eucalypt plantations established on a ferralic arenosols, Congo, *Forest Ecosyst.* 7(1), 40. DOI: 10.1186/s40663-020-00249-w.

Koutika, L. S., Fiore, A., Tabacchioni, S., Aprea, G., Pereira, A. PdA. and Bevivino, A. (2020b). Influence of *Acacia mangium* on soil fertility and bacterial community in *Eucalyptus* Plantations in the Congolese Coastal Plains, *Sustainability* 12(21), 8763. DOI: 10.3390/su12218763.

Koutika, L.-S., Taba, K., Ndongo, M. and Kaonga, M. (2021). Nitrogen-fixing trees increase organic carbon sequestration in forest and agroforestry ecosystems in the Congo basin, *Reg. Environ. Change* 21(4). DOI: 10.1007/s10113-021-01816-9.

Lescuyer, G., Karsenty, A. and Eba'a Atyi, R. (2009) A new tool for sustainable forest management in Central Africa: payments for environmental services. Chap 8, 127–139.

Lorenz, K. and Lal, R. (2005). The depth distribution of soil organic carbon in relation to land use and management and the potential of carbon sequestration in subsoil horizons, Adv. Agron. 88, 35–66.

Lorenz, K. and Lal, R. (2014). Soil organic carbon sequestration in agroforestry systems: a review, *Agron. Sustain. Dev.* 34(2), 443–454. DOI: 10.1007/s13593-014-0212-y.

Ma, Z. and Chen, H. Y. H. (2017). Effects of species diversity on fine root productivity increase with stand development and associated mechanisms in a boreal forest, *J. Ecol.* 105(1), 237–245. DOI: 10.1111/1365-2745.12667.

Ma, Z. L., Chen, H. Y. H., Bork, E. W., Carlyle, C. N. and Chang, S. X. (2020). Carbon accumulation in agroforestry systems is affected by tree species diversity, age and regional climate: a global meta-analysis, *Glob. Ecol. Biogeogr.* 29(10), 1817–1828. DOI: 10.1111/geb.13145.

Marsden, C., Martin-Chave, A., Cortet, J., Hedde, M. and Capowiez, Y. (2019). How agroforestry systems influence soil fauna and their functions - a review, *Plant Soil* 453(1–2), 29–44. DOI: 10.1007/s11104-019-04322-4.

Mayer, S., Wiesmeier, M., Sakamoto, E., Hübner, R., Cardinael, R., Kühnel, A. and Kögel-Knabner, I. (2022). Soil organic carbon sequestration in temperate agroforestry systems – a meta-analysis, *Agric. Ecosyst. Environ.* 323, 107689. DOI: 10.1016/j. agee.2021.107689.

Mbow, C., Smith, P., Skole, D., Duguma, L. and Bustamante, M. (2014). Achieving mitigation and adaptation to climate change through sustainable agroforestry practices in Africa, *Current Opinion in Environmental Sustainability* 6, 8–14. DOI: 10.1016/j. cosust.2013.09.002.

Nair, P. K. R. (1985). Classification of agroforestry systems, *Agroforest. Syst.* 3(2), 97–128. DOI: 10.1007/BF00122638.

Nair, P. K. R., Kumar, B. M. and Nair, V. D. (2009a). Agroforestry as a strategy for carbon sequestration, *J. Plant Nutr. Soil Sci.* 172, 10–23. DOI: 10.1002/jpln.200800030.

Nair, P. K. R., Nair, V. D., Kumar, B. M. and Haile, S. G. (2009b). Soil carbon sequestration in tropical agroforestry systems: a feasibility appraisal, *Environ. Sci. Policy* 12(8), 1099–1111. DOI: 10.1016/j.envsci.2009.01.010.

Nair, P. K. R., Nair, V. D., Mohan Kumar, B. and Showalter, J. M. (2010). Carbon sequestration in agroforestry systems, Adv. Agron. 108, 237–307. DOI: 10.1016/ S0065-2113(10)08005-3.

Nair, P. K. R. (2012). Carbon sequestration studies in agroforestry systems: a reality-check, *Agroforest. Syst.* 86(2), 243–253.

Nguyen, Q., Hoang, M. H., Oborn, I. and Noordwijk, M. V. (2013). Multipurpose agroforestry as a climate change resiliency option for farmers: an example of local adaptation in Vietnam, *Clim. Change* 117(1–2), 241–257.

Nsombo, B. M. (2016). Evaluation des nutriments et du carbone organique du sol dans le système agroforestier du plateau des Batéké en République Démocratique du Congo. Thèse de Doctorat. Ecole Régionale Post-Universitaire d'Aménagement et de Gestion Intégrés des forêts et Territoires Tropicaux (ERAIFT), pp. 79–80.

Nygren, P., Fernandez, M. P., Harmand, J. M. and Leblanc, H. A. (2012). Symbiotic dinitrogen fixation by trees: an underestimated resource in agroforestry systems?, *Nutr. Cycl. Agroecosyst.* 94(2–3), 123–160. DOI: 10.1007/s10705-012-9542-9.

Ortaş, İ, Lal, R. and Kapur, S. (2017). Carbon sequestration and mycorrhizae in Turkish soils, *The Anthropocene: Politik–Economics–Society–Science*, 139–149. DOI: 10.1007/978-3-319-45035-3_10.

Pandey, D. N. (2002). Carbon sequestration in agroforestry systems, *Clim. Policy* 2, 367–377.

Paula, R. R., Bouillet, J., Ocheuze Trivelin, P. C., Zeller, B., Leonardo de Moraes Gonçalves, J., Nouvellon, Y., Bouvet, J., Plassard, C. and Laclau, J. (2015). Evidence of short-term belowground transfer of nitrogen from *Acacia mangium* to *Eucalyptus grandis* trees in a tropical planted forest, *Soil Biol. Biochem.* 91, 99–108.

Pellerin, S., Bamière, L., Launay, C., Martin, R., Schiavo, M., Angers, D., Augusto, L., Balesdent, J., Basile-Doelsch, I., Bellassen, V., Cardinael, R., Cécillon, L., Ceschia, E., Chenu, C., Constantin, J., Darroussin, J., Delacote, P., Delame, N., Gastal, F., Gilbert, D., Graux, A. I., Guenet, B., Houot, S., Klumpp, K., Letort, E., Litrico, I., Martin, M., Menasseri, S., Mézière, D., Morvan, T., Mosnier, C., Roger-Estrade, J., Saint-André, L., Sierra, J., Thérond, O., Viaud, V., Grateau, R., Le Perchec, S., Savini, I. and Réchauchère, O. (2020). Stocker du carbone dans les sols français, Quel potentiel au regard de l'objectif 4 pour 1000 et à quel coût ? *Rapport scientifique de l'étude*, INRA, France, 540 p.

Pereira, A. P. A., de Andrade, P. A. M., Bini, D., Durrer, A., Robin, A., Bouillet, J. P., Andreote, F. D. and Cardoso, E. J. B. N. (2017). Shifts in the bacterial community composition along deep soil profiles in monospecific and mixed stands of *Eucalyptus grandis* and *Acacia mangium*, *PLoS ONE* 12(7), e0180371.

Pereira, A. P. A., Zagatto, M. R. G., Brandani, C. B., Mescolotti, D. L., Cotta, S. R., Gonçalves, J. L. M. and Cardoso, E. J. B. N. (2018). Acacia changes microbial indicators and increases C and N in soil organic fractions in intercropped Eucalyptus plantations, *Front. Microbiol.* 9, 655. DOI: 10.3389/fmicb.2018.00655.

Pereira, A. P. A., Santana, M. C., Zagatto, M. R. G., Brandani, C. B., Wang, J. T., Verma, J. P., Singh, B. K. and Cardoso, E. J. B. N. (2020). Nitrogen-fixing trees in mixed forest systems regulate the ecology of fungal community and phosphorus cycling, *Sci. Total Environ.* 758, 143711 DOI: 10.1016/j.scitotenv.2020.143711.

Poeplau, C., Don, A., Vesterdal, L., Leifeld, J., Van Wesemael, B., Schumacher, J. and Gensior, A. (2011). Temporal dynamics of soil organic carbon after land-use change in the temperate zone - carbon response functions as a model approach, *Glob. Change Biol.* 17(7), 2415-2427.

Prasad, R., Dhyani, S. K., Newaj, R., Kumar, S. and Tripathi, D. V. (2016). Contribution of advanced agroforestry research in sustaining soil quality for increased food production and food security, *J. Soil Water Conserv.* 15(1), 31–39.

Price, G. W. and Gordon, A. M. (1998). Spatial and temporal distribution of earthworms in a temperate intercropping system in southern Ontario, Canada, *Agroforest. Syst.* 44(2/3), 141–149.

Reich, P., Tilman, D., Forest, I., Mueller, K., Hobbie, S. E., Flynn, D. F. B. and Eisenhauer, N. (2012). Impacts of biodiversity loss escalate through time as redundancy fades, *Science* 336(6081) 589–592. DOI: 10.1126/science.1217909.

Rosenstock, T. S., Wilkes, A., Jallo, C., Namoi, N., Bulusu, M., Suber, M., Mboi, D., Mulia, R., Simelton, E., Richards, M., Gurwick, N. and Wollenberg, E. (2019). Making trees count: measurement and reporting of agroforestry in UNFCCC national communications of non-Annex I countries, *Agric. Ecosyst. Environ.* 284, 106569. DOI: 10.1016/j. agee.2019.106569.

Rumpel, C., Amiraslani, F., Chenu, C., Garcia Cardenas, M., Kaonga, M., Koutika, L. S., Ladha, J., Madari, B., Shirato, Y., Smith, P., Soudi, B., Soussana, J. F., Whitehead, D. and Wollenberg, E. (2019). The 4p1000 Initiative: opportunities, limitations and challenges for implementing soil organic carbon sequestration as a sustainable development strategy, *Ambio* 49(1), 350–360.

Sanginga, N., Mulungoy, K. and Ayanaba, A. (1986). Inoculation of *Leucaena leucocephala* Lam de Witt with Rhizobium and its nitrogen contribution to a subsequent maize crop, *Biol. Agric. Hortic.* 3, 341–352.

Santos, F. M., Chaer, G. M., Diniz, A. R. and Balieiro, FdC. (2017a). Nutrient cycling over five years of mixed-species plantations of Eucalyptus and Acacia on a sandy tropical soil, *Forest Ecol. Manag.* 384, 110–121. DOI: 10.1016/j.foreco.2016.10.041.

Santos, F. M., Balieiro, FdC., Fontes, M. A. and Chaer, G. M. (2017b). Understanding the enhanced litter decomposition of mixed-species plantations of *Eucalyptus* and *Acacia mangium, Plant Soil* 423(1–2), 141–155. DOI: 10.1007/s11104-017-3491-7.

Schroth, G. and Sinclair, F. L. (2003). Chapter 1: impacts of trees on the fertility of agricultural soils. In: *Trees, Crops and Soil Fertility. Concepts and Research Methods* Schroth, G. and Sinclair, F. L. (Eds), CAB International; Oxford University Press, UK, pp. 1–13.

Sebukyu, V. B. and Mosango, M. (2012). Adoption of agroforestry systems by farmers in Masaka District of Uganda, *Ethnobot. Res. Appl.* 10, 059–068. Available at: ethno botanyjournal.org/vol10/i1547-3465-10-059.pdf.

Shi, L. L., Feng, W. T., Xu, J. C. and Kuzyakov, Y. (2018). Agroforestry systems: meta-analysis of soil carbon stocks, sequestration processes, and future potentials, *Land Degrad. Dev.* 29(11), 3886–3897. DOI: 10.1002/ldr.3136.

Sileshi, G. and Mafongoya, P. L. (2006). Long-term effects of improved legume fallows on soil invertebrate macrofauna and maize yield in eastern Zambia, *Agric. Ecosyst. Environ.* 115(1–4), 69–78. DOI: 10.1016/j.agee.2005.12.010.

Sileshi, G., Akinnifesi, F. K., Ajayi, O. C., Chakeredza, S., Kaonga, M. and Matakala, P. W. (2007). Contributions of agroforestry to ecosystem services in the Miombo eco-region of eastern and southern Africa, *Afr. J. Environ. Sci. Technol.* 1(4), 68–80.

Sileshi, G., Akinnifesi, F. K., Ajayi, O. C. and Place, F. (2008). Meta-analysis of maize yield response to woody and herbaceous legumes in the sub-Saharan Africa, *Plant Soil* 307(1–2), 1–19.

Six, J., Elliott, E. T. and Paustian, K. (2000). Soil macroaggregate turnover and microaggregate formation: a mechanism for C sequestration under no-tillage agriculture, *Soil Biol. Biochem.* 32(14), 2099–2103.

Somarriba, E. (1992). Revisiting the past - an essay on agroforestry definition, *Agroforest. Syst.* 19(3), 233–240.

Soto-Pinto, L., Anzueto, M., Mendoza, J., Ferrer, G. J. and de Jong, B. (2010). Carbon sequestration through agroforestry in indigenous communities of Chiapas, Mexico, *Agroforest. Syst.* 78(1), 39–51.

Stainback, G. A., Masozera, M., Mukuralinda, A. and Dwivedi, P. (2012). Smallholder agroforestry in Rwanda: a SWOT-AHP analysis, *Small-Scale For.* 11(3), 285–300. DOI: 10.1007/s11842-011-9184-9.

Stavi, I. and Lal, R. (2013). Agroforestry and biochar to offset climate change: a review, *Agron. Sustain. Dev.* 33(1), 81–96. DOI: 10.1007/s13593-012-0081-1.

Thomas, A. L., Kallenbach, R., Sauer, T. J., Brauer, D. K., Burner, D. M., Coggeshall, M. V., Dold, C., Rogers, W., Bardhan, S. and Jose, S. (2020). Carbon and nitrogen accumulation within four black walnut alley cropping sites across Missouri and Arkansas, USA. *Agrofor. Syst.* 94(5), 1625–1638.

Torquebiau, E. F. (2000). A renewed perspective on agroforestry concepts and classification, *C. R. Acad. Sci. III* 323(11), 1009–1017.

Udawatta, R. P., Kremer, R. J., Adamson, B. W. and Anderson, S. H. (2008). Variations in soil aggregate stability and enzyme activities in a temperate agroforestry practice, *Appl. Soil Ecol.* 39(2), 153–160.

Van Groningen, J. W., van Kessel, C., Hungate, B. A., Oenema, O., Powlson, D. S. and van Groenigen, K. J. (2017). Sequestering soil organic carbon: a nitrogen Dilemma, *Environ. Sci. Technol.* 51(9), 4738–4739. DOI: 10.1021/acs.est.7b01427.

van Noordwijk, M., Coe, R., Sinclair, F. L., Luedeling, E., Bayala, J., Muthuri, C. W., Cooper, P., Kindt, R., Duguma, L., Lamanna, C. and Minang, P. A. (2021). Climate change adaptation in and through agroforestry: four decades of research initiated by Peter Huxley, *Mitig Adapt. Strateg. Glob. Chang* 26(5), 1–33. DOI: 10.1007/s11027-021-09954-5.

Verchot, L. V., Van Noordwijk, M., Kandji, S., Tomich, T., Ong, C., Albrecht, A., Mackensen, J., Bantilan, C., Anupama, K. V. and Palm, C. (2007). Climate change: linking adaptation and mitigation through agroforestry, *Mitig Adapt. Strat. Glob. Change* 12(5), 901–918. DOI: 10.1007/s11027-007-9105-6.

von Haden, A. C., Kucharik, C. J., Jackson, R. D. and Marin-Spiotta, E. (2019). Litter quantity, litter chemistry, and soil texture control changes in soil organic carbon fractions under bioenergy cropping systems of the North Central US, *Biogeochemistry* 143(3), 313–326. DOI: 10.1007/s10533-019-00564-7.

Wang, Z. G., Bi, Y. L., Jiang, B., Zhakypbek, Y., Peng, S. P., Liu, W. W. and Liu, H. (2017). Arbuscular mycorrhizal fungi enhance soil carbon sequestration in the coalfields, northwest China, Nature Scient. Rep. 6, 34336. DOI: 10.1038/srep34336.

Zomer, R., Trabucco , C. R., Place, F., van Noordwijk, M. and Xu, J. (2014). Trees on farms: an update and reanalysis of agroforestry's global extent and socio-ecological characteristics, *Working Paper 179. Bogor, Indonesia: World Agroforestry Centre (ICRAF) Southeast Asia Regional Program.* DOI: 10.5716/WP14064.PDF.

Zomer, R. J., Neufeldt, H., Xu, J., Ahrends, A., Bossio, D., Trabucco, A., Van Noordwijk, M. and Wang, M. (2016). Global tree cover and biomass carbon on agricultural land: the contribution of agroforestry to global and national carbon budgets, *Sci. Rep.* 6, 29987. DOI: 10.1038/srep29987.

Chapter 3

Advances in monitoring and reporting forest emissions and removals in the context of the United Nations Framework Convention on Climate Change (UNFCCC)[1,2]

Marieke Sandker and Till Neeff, Food and Agriculture Organization of the United Nations (FAO), Italy

1 Introduction

1.1 REDD+ under the UNFCCC and sustainable forest management

The landmark Paris Agreement on climate change, adopted by the 21st Conference of the Parties to the United Nations Framework Convention on Climate Change in 2015, recognizes forests as important sources and sinks of greenhouse gases. It includes reducing emissions from deforestation and forest degradation (REDD+) as one of the strategies for increasing the carbon-sink capacity of forests and reducing their emissions.

The world's tropical forests store vast amounts of carbon. Conserving and sustainably managing these forests, therefore, must be a global priority, not only for achieving sustainable development goal (SDG) 15 ('sustainably manage forests, combat desertification, halt and reverse land degradation,

1 This chapter is closely based on FAO's recent Forestry Working Paper 9 *From Reference Levels to Results Reporting: REDD+ Under the United Nations Framework Convention on Climate Change*. Available at: http://www.fao.org/d ocuments/card/en/c/ca6031en. This publication was made possible through support from the UN-REDD Programme at FAO, with financial contributions from the governments of Denmark, Japan, Luxembourg, Norway, Spain, Switzerland and the European Union.
2 The views expressed in this publication are those of the author(s) and do not necessarily reflect the views or policies of the Food and Agriculture Organization of the United Nations.

http://dx.doi.org/10.19103/AS.2020.0074.28
Published by Burleigh Dodds Science Publishing Limited, 2021.

and halt biodiversity loss'), but crucially also for SDG 13 ('take urgent action to combat climate change and its impacts').

Many countries are making progress in developing REDD+ National Strategies and/or Action Plans, developing and submitting REDD+ Forest Reference (Emission) Levels (FREL/FRLs) and submitted REDD+ results to the United Nations Framework Convention on Climate Change (UNFCCC) for technical assessment.

1.2 Forest monitoring and reporting for REDD+ through FREL/FRL and REDD+ results submissions

Countries use their national forest monitoring systems (NFMS) to measure REDD+ results, mostly including data from national forest inventories and satellite land-monitoring systems. Countries report on REDD+ results through FREL/FRL submissions and the REDD+ results' annexes (the latter are contained in dedicated annexes to Biennial Update Reports (BUR)). These undergo technical assessments and technical analyses.

Countries voluntarily submit FREL/FRLs to the UNFCCC for technical assessment (TA). In doing so, they may apply for results-based payments under the financing mechanism of the UNFCCC (Green Climate Fund). The UNFCCC provides guidelines and modalities through its Conference of the Parties (COP) decisions for FREL/FRL construction (Fig. 1). The TA will evaluate the extent to which the FREL/FRL submission is in line with the guidelines contained in the relevant COP decisions. Once a TA has been completed, countries can submit REDD+ results in an annex to their BURs for technical analysis (Fig. 3).

Preceding FAO publications (2015a, 2017, 2018a) provide more detailed explanation of UNFCCC guidance and modalities for FREL/FRL and REDD+ results submissions.

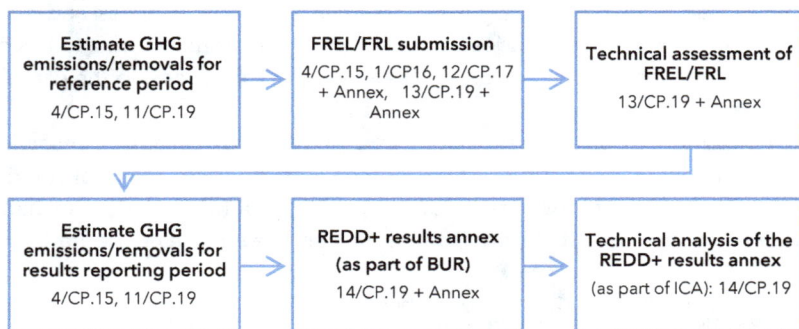

Figure 1 Measurement, reporting and verification for REDD+, and the most relevant decisions of the UNFCCC.

1.3 Status of FREL/FRL and REDD+ results submissions

As of March 2019, 39 countries had submitted 45 FREL/FRLs to the UNFCCC,[2] comprising 12 countries in Africa, 13 in Asia and the Pacific and 14 in Latin America and the Caribbean (Fig. 2). Collectively, the countries that submitted their FREL/FRLs to the UNFCCC are home to a forest area of approximately 1.49 billion ha (37% of the global forest area) and contribute to around 70% of global forest area loss.[3] Those countries that submitted REDD+ results to the UNFCCC subsequent to the FREL/FRL submission, collectively account for a forest area of 744 million ha (19% of the global forest area) and total around 35% of global forest area loss. For the 45 FREL/FRL submissions, 37 TA reports had been published by June 2019.

Eleven[4] submissions of REDD+ results were included in the technical annexes of the BURs of eight countries. Of these, technical analyses were completed on five submissions by early July 2019 as part of the international consultation and analysis (ICA) process (Fig. 3).

Four countries have submitted FREL/FRLs more than once (see Section 2.1). Brazil and Colombia submitted more than one technical annex with REDD+ results. Brazil submitted results in its biennial update report (BUR) 1, 2 and 3 for three reporting periods for the Amazon region, and results for the Cerrado region in BUR 3. Colombia submitted results for two reporting periods for the Colombian Amazon.

2 Summary of UNFCCC FREL/FRLs

2.1 What's new from FREL/FRL submissions

In 2019, an additional seven countries submitted an FREL/FRL to the UNFCCC. Five of these submitted for the first time (Argentina, Bangladesh, Guinea-Bissau, Nicaragua and Solomon Islands) whereas two countries submitted for the second and third time, Nigeria and Malaysia, respectively.

Submissions show increased reporting of uncertainty around activity data, continuing the trend from previous years (Fig. 4). Out of the seven new submissions, five included an estimate of aggregate uncertainty of the FREL/FRL (see Section 2.3).

---

2 The 39 countries comprised of 25% of the UNFCCC's 154 non-Annex I countries (Annex 1 includes the industrialized countries that were members of the OECD (Organisation for Economic Co-operation and Development) in 1992, plus countries with economies in transition (EIT Parties), including the Russian Federation, the Baltic States, and several Central and Eastern European countries).

3 Forest area and forest area loss estimates are derived from FRA2015 (FAO, 2015c); the estimates are not based on forest area or deforestation estimates in the FREL/FRL submissions since these are not consistently reported in the FREL/FRL submissions. For subnational FREL/FRL submissions the national areas are considered. Global forest area loss refers to the sum of net loss of forest area at country level, to which the FREL/FRL and REDD+ results-submitting countries contributed 73% and 39%, respectively, in 1990-2015, and 69% and 34%, respectively, in 2000-2015.

4 Brazil's latest BUR contains a technical annex with REDD+ results for the Amazon (2016-2017) and a technical annex with REDD+ results for the Cerrado (2011-2017), which here is considered as one REDD+ results submission.

Published by Burleigh Dodds Science Publishing Limited, 2021.

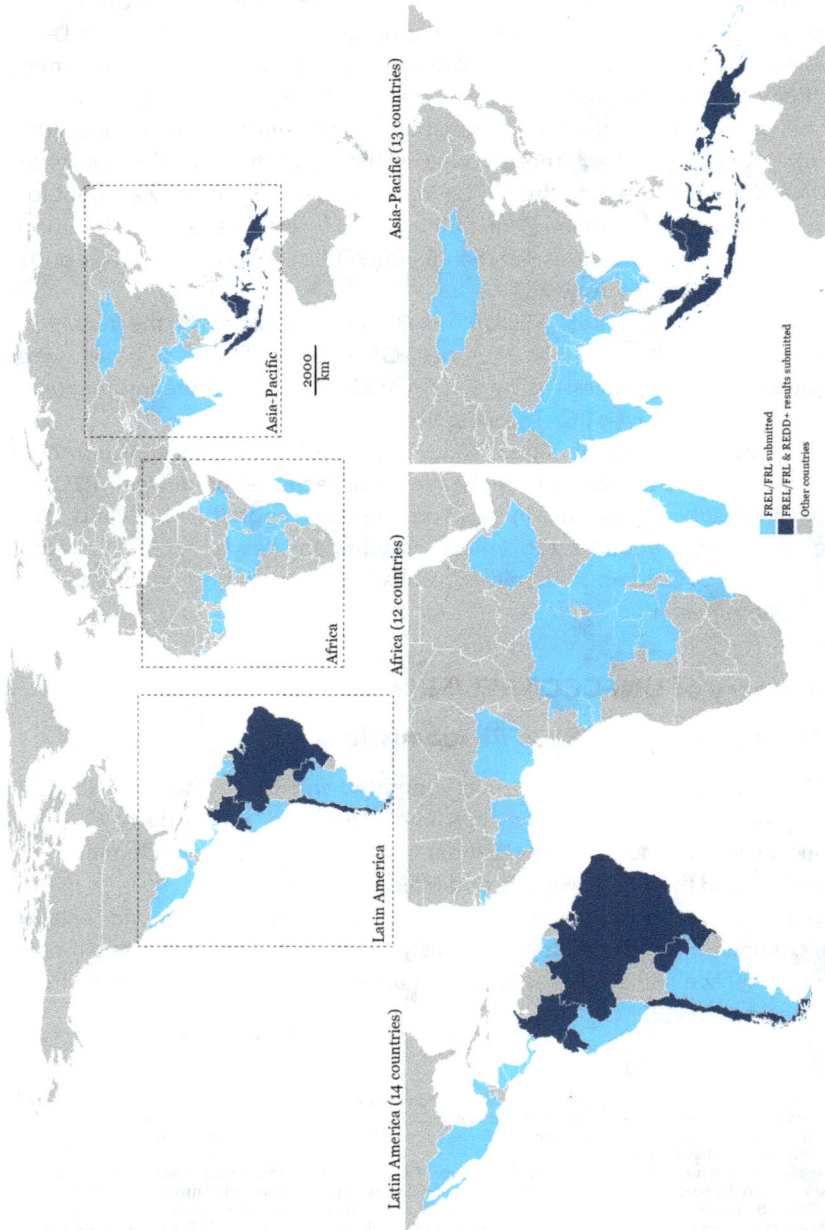

Figure 2 Geographical distribution of countries that have submitted a FREL/FRL (light blue) and those that subsequently submitted REDD+ results (dark blue) to the UNFCCC.

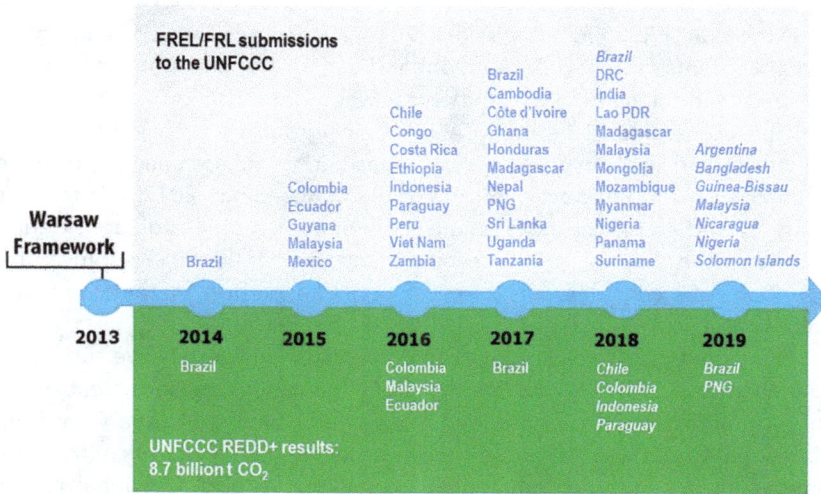

Figure 3 Overview of FREL/FRL and REDD+ results submissions to the UNFCCC. Notes: COP19, held in November 2013 in Warsaw, Poland, adopted the seven decisions of the Warsaw Framework for REDD-plus (UNFCCC, 2013). Country names in italic indicate that their TA was ongoing by September 2019. Brazil's 2019 BUR includes two REDD+ results technical annexes, one for Amazon C and one for Cerrado. DRC = Democratic Republic of the Congo; Lao PDR = Lao People's Democratic Republic; PNG = Papua New Guinea.

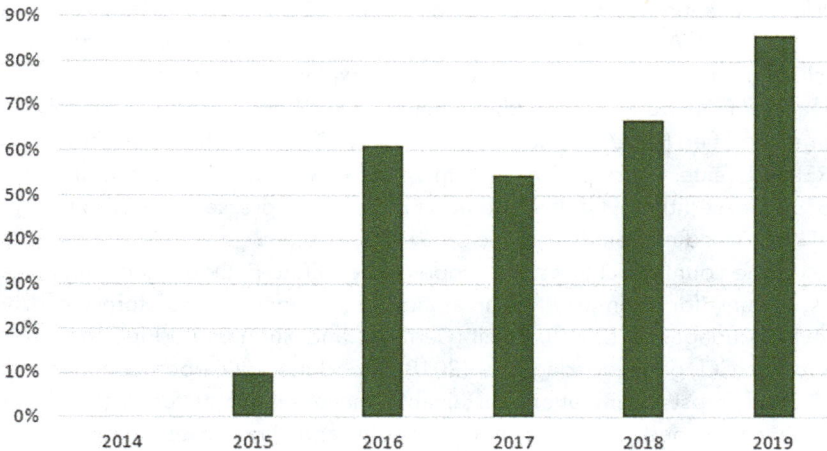

Figure 4 Percentage of submissions that include uncertainty estimates (i.e. confidence intervals) around activity data used for FREL/FRLs. Note: Some countries provided uncertainty estimates without actually calculating them. For example, one country estimated uncertainty and another derived an uncertainty estimate based on overall map accuracy (which is not necessarily a good estimator for AD uncertainty (Olofsson et al., 2013).

In the initial FREL/FRL submissions (2014–2016), no country considered the carbon contents in post-deforestation land use for the emission estimations. They assumed full carbon loss in the year when deforestation was measured from the aboveground biomass (AGB), belowground biomass (BGB), litter and deadwood carbon pools without considering subsequent removals in vegetation growing in post-deforestation land use. Since 2017, this trend is changing and technical assessments commend countries for considering the carbon content in post-deforestation land use. Of the 2019 submissions, 57% took into consideration post-deforestation carbon in their emission estimations.

Four countries (Brazil, Madagascar, Malaysia and Nigeria) have submitted more than one FREL/FRL to the UNFCCC, for various reasons: to expand the geographic scope (Brazil added the Cerrado biome, Nigeria went from subnational to national coverage); to cover more REDD+ activities (Malaysia added conservation and reduced deforestation); to add additional results reporting period (Brazil added 'Amazon C' for results in 2016–2020); and to update the FREL with new, improved data and updated reference period (Madagascar).

2.2 Trends in the technical assessment

Almost all countries submitted a modified FREL/FRL during the technical assessment (TA). Sometimes these modified FREL/FRL submissions only include more and better information (i.e. explaining more clearly how the measurements were made) without changing the FREL/FRL value, but for 81% of the modified FREL/FRL submissions the FREL/FRL value has changed. Of the FREL/FRL values expressed in (net) emissions, 46% reduced net emissions while 35% increased them. Of the modified FRL values expressed in (net) removals, 45% increased net removals while 36% reduced them.

Some countries change the scope of their FREL/FRL during the technical assessment, for example, Myanmar added enhancement of forest carbon stocks; Madagascar (2017), Brazil (Cerrado) and Suriname added emissions from non-CO_2 gases; Madagascar (2018) added the soil carbon pool; Uganda removed forest degradation, sustainable management of forest (SMF) and conservation; and Chile, Guyana, Mongolia and Panama removed the soil carbon pool as a result of the TA. The most common reason for omissions of activities or carbon pools during the TA is related to concerns around the accuracy and reliability of the data.

The TA reports frequently highlight the assessment of uncertainty as an area for improvement and more in-depth assessments of uncertainty are mentioned over time, including capturing all sources of error and using higher approaches (e.g. Monte Carlo simulation).

Published by Burleigh Dodds Science Publishing Limited, 2021.

The TA may also comment on the FREL/FRL construction approach and reference period. For example, as a result of their technical assessments, Malaysia changed its reference period, Ghana changed its initially proposed linear projection FREL to a historical average, and Myanmar substituted a zero FRL for enhancement with average removals over the reference period.

2.3 Choices made by countries on FREL/FRL elements

The FAO (2018a) provides a comprehensive overview of UNFCCC guidance on each of the FREL/FRL elements, an overview of country choices and in some cases an explanation of why countries made these choices, and a summary of responses that countries received from the technical assessment. This publication provides a brief summary of country choices per FREL/FRL element, in some cases illustrated with examples from new FREL/FRL submissions.

2.3.1 Forest definition

Most countries included references to threshold parameters and the use of the land (Fig. 5). The majority choose a minimum area threshold of 0.5 ha or 1 ha, a minimum height of 5 m, and a minimum canopy cover of either 10% or 30%. Several countries use the FAO's Global Forest Resources Assessment (FRA) thresholds: a canopy cover of 10%; a tree height of 5 m; and an area of 0.5 ha.

Some countries diverge from the forest definition they adopted for REDD+ and use an operational forest definition, generally because of technical limitations with their measurement, reporting and verification. For example, Indonesia uses a legal definition for management purposes of 0.25 ha but a working definition for MRV of 6.25 ha. Similar differences in minimum area apply to Brazil and Nepal, which both have 0.5 ha in their forest definitions, but apply 6.25 ha and 2.25 ha, respectively, due to technical limitations in the (historical) data. In view of possible displacement of emissions, the TA asked these countries to monitor small-scale deforestation, as large-scale deforestation can be reduced whereas small-scale deforestation may increase. Another recurring area for technical improvement is the exclusion of temporarily unstocked forest land in the deforestation area estimate.

2.3.2 Scale

Most FREL/FRL submissions (80%) are national scale (Fig. 6). Of the nine subnational submissions, seven are from Latin American countries and two from African countries. One country (Nigeria) first submitted a subnational FREL and subsequently a national FREL.

Minimum area threshold in forest definition

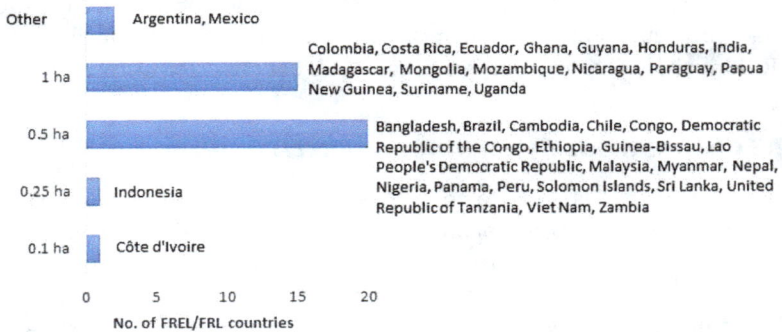

Other — Argentina, Mexico

1 ha — Colombia, Costa Rica, Ecuador, Ghana, Guyana, Honduras, India, Madagascar, Mongolia, Mozambique, Nicaragua, Paraguay, Papua New Guinea, Suriname, Uganda

0.5 ha — Bangladesh, Brazil, Cambodia, Chile, Congo, Democratic Republic of the Congo, Ethiopia, Guinea-Bissau, Lao People's Democratic Republic, Malaysia, Myanmar, Nepal, Nigeria, Panama, Peru, Solomon Islands, Sri Lanka, United Republic of Tanzania, Viet Nam, Zambia

0.25 ha — Indonesia

0.1 ha — Côte d'Ivoire

No. of FREL/FRL countries

Height threshold in forest definition

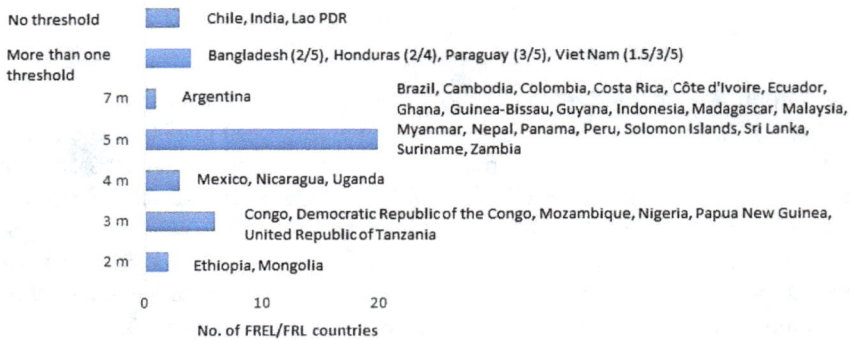

No threshold — Chile, India, Lao PDR

More than one threshold — Bangladesh (2/5), Honduras (2/4), Paraguay (3/5), Viet Nam (1.5/3/5)

7 m — Argentina

5 m — Brazil, Cambodia, Colombia, Costa Rica, Côte d'Ivoire, Ecuador, Ghana, Guinea-Bissau, Guyana, Indonesia, Madagascar, Malaysia, Myanmar, Nepal, Panama, Peru, Solomon Islands, Sri Lanka, Suriname, Zambia

4 m — Mexico, Nicaragua, Uganda

3 m — Congo, Democratic Republic of the Congo, Mozambique, Nigeria, Papua New Guinea, United Republic of Tanzania

2 m — Ethiopia, Mongolia

No. of FREL/FRL countries

Canopy cover threshold in forest definition

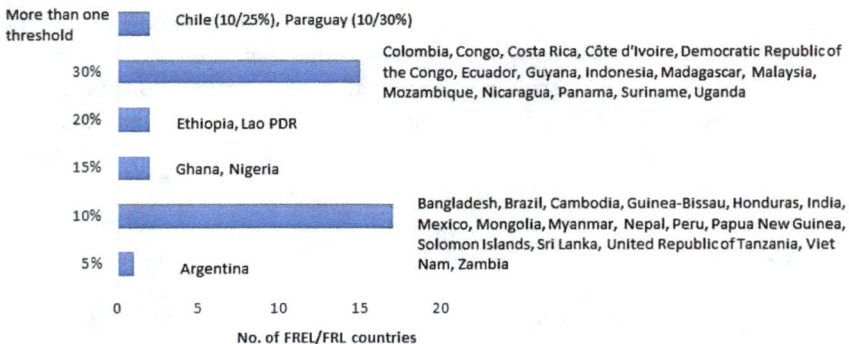

More than one threshold — Chile (10/25%), Paraguay (10/30%)

30% — Colombia, Congo, Costa Rica, Côte d'Ivoire, Democratic Republic of the Congo, Ecuador, Guyana, Indonesia, Madagascar, Malaysia, Mozambique, Nicaragua, Panama, Suriname, Uganda

20% — Ethiopia, Lao PDR

15% — Ghana, Nigeria

10% — Bangladesh, Brazil, Cambodia, Guinea-Bissau, Honduras, India, Mexico, Mongolia, Myanmar, Nepal, Peru, Papua New Guinea, Solomon Islands, Sri Lanka, United Republic of Tanzania, Viet Nam, Zambia

5% — Argentina

No. of FREL/FRL countries

Figure 5 Threshold values for REDD+ forest definitions, by country.

Published by Burleigh Dodds Science Publishing Limited, 2021.

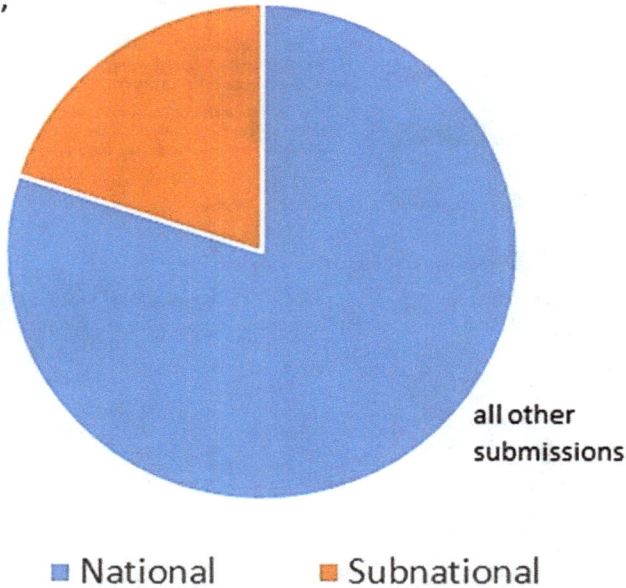

Figure 6 Scale of FREL/FRL submissions. Note: All other submissions include Nigeria, which is the only country to date that has submitted both a subnational and a national FREL.

For some countries, national FREL/FRLs comprised the sum of subnational jurisdictional FRELs/FRLs, such as provinces (e.g. in the cases of the Democratic Republic of the Congo and Madagascar). Other countries (e.g. Zambia) indicated their intention to disaggregate national estimates in the future. Such disaggregation would provide countries with information on performance at the subnational jurisdictional level, which may be of particular value if jurisdictions move at different speeds in REDD+ implementation.

2.3.3 Scope: REDD+ activities, pools and gases included

Concerning the scope of REDD+ activities, deforestation remains the most frequently included REDD+ activity in FREL/FRL submissions, with 96% of the submissions including the activity (Fig. 7).

Several countries submit the FRL with activities such as deforestation, forest degradation and enhancement indicating that all greenhouse gas (GHG) fluxes from the forest have been covered. Lee, Skutsch and Sandker (2018a) explain how the 'plus' activities of enhancement, sustainable management of forest

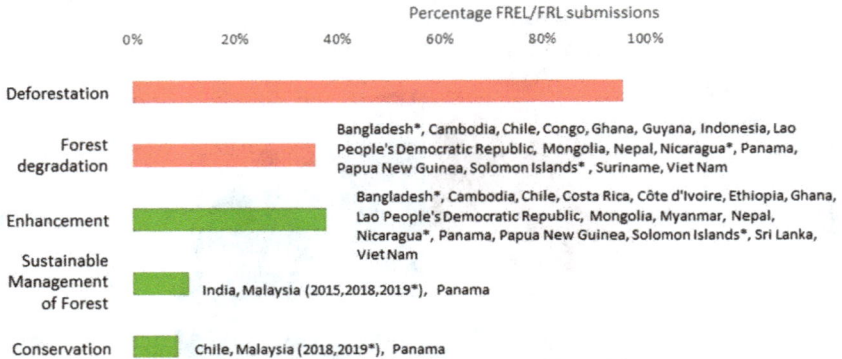

Figure 7 REDD+ activities included in FREL/FRL submissions. Notes: With the exception of Malaysia (2015) and India, all submissions included deforestation. *Countries with ongoing TAs; scope may still change.

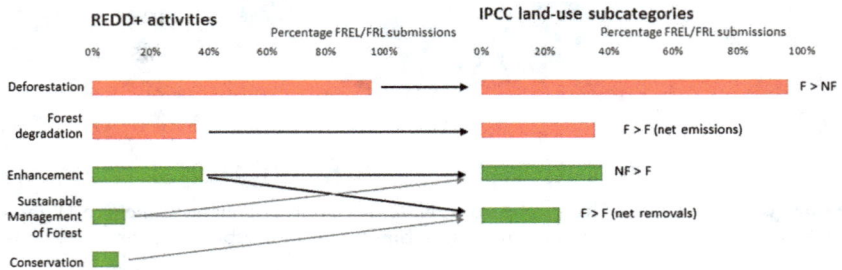

Figure 8 How the REDD+ activities included in the FREL/FRL submissions correspond to IPCC land-use subcategories. Note: Chile's conservation also includes emissions from forest degradation in conservation areas, but the activity still results in net removals. Likewise, India's SMF includes afforestation, harvesting (deforestation and forest degradation), thinning (forest degradation) and forest conservation as a management practice.

(SMF) and conservation of forest carbon stocks were added in the negotiations to ensure that a wide range of activities were covered by REDD+, without fully considering how this would be reported. Figure 8 shows the overlap among the 'plus' activities, where the same carbon fluxes are named differently by different countries.

Concerning the scope of carbon pools, above- and below-ground biomass remain the most frequently reported carbon pools in the FREL/FRL submissions.

Deadwood is mostly included by countries that assessed this pool in their national forest inventory (NFI). The Intergovernmental Panel on Climate Change (IPCC) 2006 guidelines do not provide default values for deadwood indicating there was too little coherence in the literature to propose a single value.

Published by Burleigh Dodds Science Publishing Limited, 2021.

Litter is equally included by countries that assessed their pool in their NFI (often only measured in a subset of plots) or used IPCC default values.

To date, five countries have included the soil carbon pool: Ghana, India, Indonesia, Madagascar and Malaysia. Soil emissions from mineral soils are included by countries for deforestation, afforestation[5] or both. Emissions from organic soils, through peat drainage, are included for deforestation and in Indonesia's submission, also, for forest degradation. According to IPCC's Tier 1 approach, mineral soils are in equilibrium for forest land remaining forest land (FL–FL), meaning forest degradation on mineral soils can be assumed not to cause emissions from the soil carbon pool. Countries estimate soil carbon either using national data (e.g. India assessed soil carbon contents with its NFI[6]) or IPCC default factors (e.g. Ghana). Several countries omitted the soil carbon pool after they found it not to be significant (e.g. Nigeria's 2019 submissions assessed it as <10% of emissions from the biomass pools). However, most countries omitting soil carbon indicate that they have too little data to estimate it. Even an IPCC default calculation, for example, requires information on the replacing land use and management regime applied. Some countries also mention challenges in reporting soil emissions as an argument for its omission, as soil emissions occur over a longer period (flowing over from the reference period into the results reporting period). Indonesia and Malaysia, the only two countries including emissions from peat drainage, do not include emissions from mineral soils. Malaysia only includes inherited emissions from cleared forest on peatlands drained during the 1960s and 1970s, claiming that no drainage is currently occurring.

Ghana included soil emissions using IPCC default values (equation 2.25, IPCC, 2006) applying a 'committed emissions' approach, meaning that it assumes all emissions to occur at the time of conversion. Ghana explains that, in the context of reference levels, delayed emissions (as the IPCC requires) lead to errors because the reference period is 15 years. A historical average of emissions over the reference period would therefore underestimate emissions, as only 5–75% of the total emissions over 20 years would be included. For this reason, Ghana uses the committed emissions approach for the reference level, arguing that it more accurately reflects the reality on the ground. The AT does not include an area for technical improvement related to the soil emission estimation for Ghana.

All submissions include CO_2 and 20% of the submissions include non-CO_2 emissions, mostly from fire, but also from drainage of peatland (Malaysia). Submissions that included non-CO_2 emissions are: Brazil (Cerrado), Chile,

---

5 India includes non-forest converted to forest land under sustainable management of forest.
6 The AT comments, however, that the use of a single value for soil carbon in forest overestimated the removals for newly planted forest over the reference period.

Costa Rica, Ghana, Madagascar (2017 and 2018 submissions), Malaysia (2018 and 2019 submissions) and Panama (Fig. 9).

2.3.4 Data selection for activity data

For deforestation (and in some cases afforestation), countries used three methods for generating activity data: (1) areas extracted directly from wall-to-wall change maps (referred to as pixel counts); (2) areas from samples that are stratified using wall-to-wall maps (referred to as stratified area estimates and described by Olofsson et al., 2014); and (3) areas from systematic sampling. These methods and their differences are explained in detail by the FAO (2018a).

Figure 10 shows that stratified area estimation and systematic sampling are becoming more common, whereas initial submissions relied on pixel counts

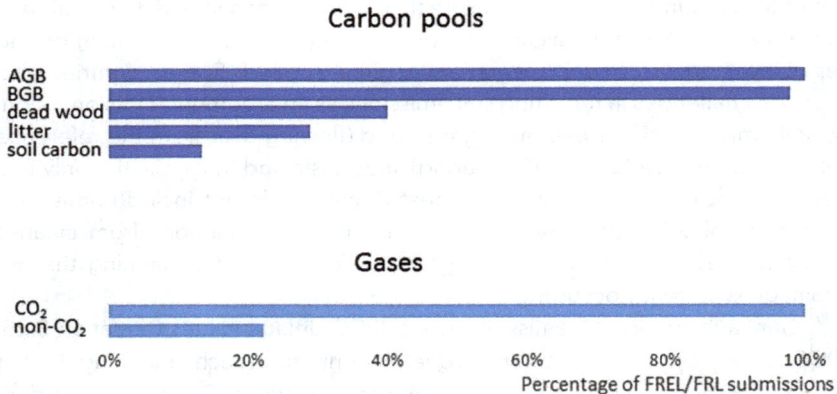

Figure 9 Scope of carbon pools and gases chosen by countries for their FREL/FRL submissions.

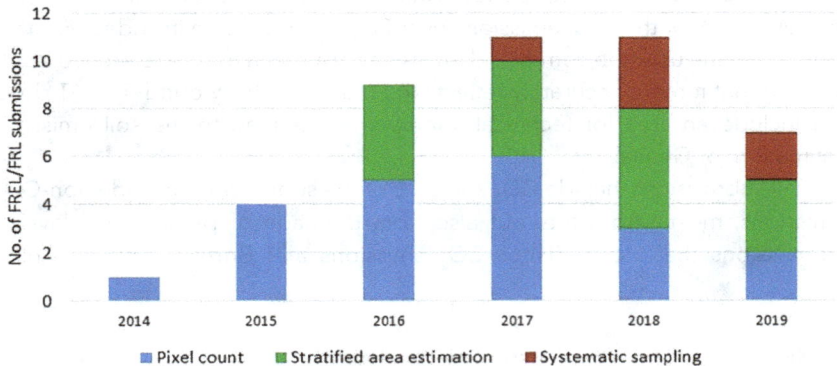

Figure 10 Methods used to assess deforestation (and in some cases afforestation also).

only. Assessing and reporting uncertainty is a requirement for participating in the Green Climate Fund's results-based payment pilot programme. Pixel counts do not allow for an assessment of the associated confidence interval around the deforestation estimate (GFOI, 2016, Section 5.1.5). Systematic sampling is sometimes used in countries where mapping approaches are not well-developed or where the country intends to obtain estimates for multiple land-use categories.

Stratified area estimation is a relatively new approach to area estimation. New findings underline the importance of the quality of reference data (sample points, which also applies to systematic sampling) and the distribution of reference data over the map strata.

For forest degradation, 36% of the FREL/FRL submissions included this activity (Fig. 7). Countries propose a variety of methods for generating activity data (Table 1). The choice of method may depend on the type of degradation and data availability. The high rate of omissions of this activity is in most cases said to be due to the lack of reliable and accurate data.

For enhancement of forest carbon stocks, 38% of the FREL/FRL submissions included this activity (Fig. 7), all those submissions included afforestation but only a few included enhancement of carbon stocks in FL-FL. For afforestation, activity data come either from satellite imagery analysis (samples, maps or a combination) or from official data on planted areas. Myanmar used data from a plantation database managed by the forest department, pointing out the challenges in identifying forest gain with remote-sensing technologies.

There are several challenges associated with assessing removal results from carbon stock enhancement against an FRL, mainly – but not only – related to the UNFCCC requirement to use historical data in the FRL and the delayed

Table 1 Methodologies proposed in FREL/FRL submissions for assessing forest degradation

Methodology	Country
Combination of remote sensing and ground inventories	Cambodia, Chile, Indonesia, Lao People's Democratic Republic, Vietnam
Multiple national forest inventory cycles	Viet Nam
Stump counts from national forest inventories	Lao People's Democratic Republic
Official timber extraction statistics	Congo, Ghana, Guyana, Suriname
Sample data interpretation of disturbance or changes in forest subdivisions and ground inventories	Mongolia, Nicaragua*, Panama, Papua New Guinea, Solomon Islands
Modelling supply/demand balance (WISDOM)	Ghana, Nepal
Proxy statistics (monitoring log truck numbers)	Ghana
MODIS (satellite sensors) burned area and IPCC default values	Ghana, Chile

Notes: Sixteen submissions (out of 45) included forest degradation. *Country with ongoing TAs; scope may still change.

removals resulting from growth. These challenges are explained in more detail in the work by Lee, Skutsch and Sandker (2018a).

For conservation of forest carbon stocks and sustainable management of forest, 9% and 11%, respectively, of the submissions included these activities. Countries that included these activities tend to report net removals (often consisting of emissions and removals) in areas subject to conservation and SMF. Basically, these countries assess the forest carbon fluxes and consequently consider additional administrative information to determine what REDD+ activity it corresponds to. For example, if the flux is happening in a national park it may be considered conservation, or if in a production forest concessions it may be considered SMF. As explained above, various other countries explain that these carbon fluxes are already covered under enhancement (and to some extent forest degradation and deforestation).

2.3.5 Data selection for emission factors

For deforestation, countries mainly use inventory data to estimate the associated emission factor (EF), either from the national forest inventory (NFI) or from local inventories. Of the 39 countries that have submitted one or more FREL/FRL(s), 56% had completed at least one NFI cycle and 28% were implementing an NFI at the time of the latest FREL/FRL submission (Fig. 11).

Most countries, in accounting for post-deforestation carbon contents, subtract the average carbon contents in replacing land-use from the average forest carbon stock of the forest type that is being deforested, but some (e.g. Côte d'Ivoire and Papua New Guinea) apply a growth rate to post-deforestation carbon stock. This may be considered in line with the IPCC guidelines, as the emissions from deforestation mainly occur in the year of conversion, but the replacing land-use carbon stock grows over several years. This does pose

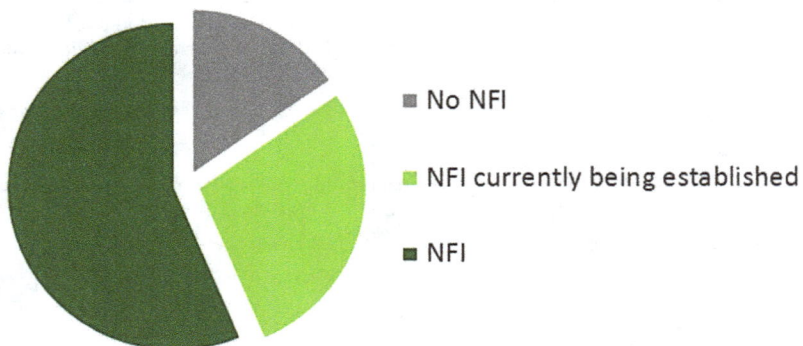

■ No NFI

■ NFI currently being established

■ NFI

Figure 11 Share of countries submitting a FREL/FRL that had undertaken or were establishing an NFI.

Published by Burleigh Dodds Science Publishing Limited, 2021.

some challenges on carbon accounting. In the case of Côte d'Ivoire, the delay in growing removals in post-deforestation land use causes the net FREL value to drop from 41 million tCO_2 in the first year of the reference period to 21 million tCO_2 in the last year. This decrease is almost entirely due to growth in post-deforestation land use (mostly perennial crops), as the deforestation area estimate does not change over the reference period.

For forest degradation, the data needed to estimate emissions largely depend on the type of activity taking place. Several countries assessing AD through (high-resolution) satellite imagery approximate the associated emissions with the difference in average carbon stock of intact and degraded/disturbed forest. Countries that consider logging to result in forest degradation sometimes use harvested volumes or other logging statistics to estimate emissions from forest degradation. In some cases, countries assess collateral damage from timber extraction (e.g. as assessed in Pearson et al., 2014) to approximate the emissions from the activity.

For enhancement, this can occur in FL–FL and other land uses converted to forest land (afforestation/reforestation).

For estimating the enhancement of forest carbon stocks in forest land remaining forest land, countries estimated removals by the difference in average carbon stock of forest types (e.g. between open and dense forest), used data on age structure applying growth models, or used data from multiple NFI cycles.

For estimating removals associated with afforestation/reforestation, some countries have applied either country-specific increment values from the NFI, in-country studies, or IPCC default growth rates. Several countries have proposed 'committed' removals, where all expected future removals are accounted for the year that afforestation was detected, which was subsequently included as an area for technical improvement in the TA reports.

2.3.6 Uncertainty analysis

Of the 2019 FREL/FRL submissions, six out of seven submissions included an uncertainty estimate around AD (Fig. 12) and six out of seven around EF, while five out of seven submissions included an aggregate uncertainty estimate around the submitted FREL/FRL value, meaning that they combined AD and EF uncertainties. For all three types of uncertainty reporting (AD, EF and aggregate) the frequency with which countries included this for 2019 is above the average percentages of uncertainties reported for all FREL/FRL submissions to date, but by far the greatest progress is seen in reporting on aggregate uncertainties, which was included in 40% of the total 45 FREL/FRL submissions and 71% of the 2019 FREL/FRL submissions (and this percentage could increase with the modified submissions). The aggregate uncertainties around the FREL/FRL

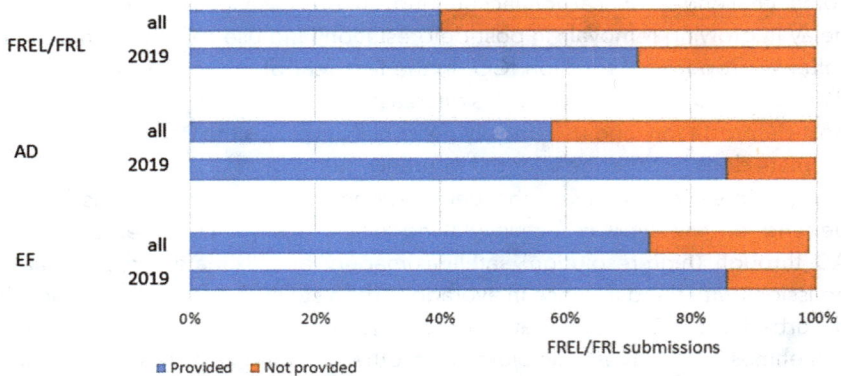

Figure 12 Percentage of submissions that provide uncertainty estimates around emission factors, activity data and overall emissions in FREL/FRLs for all 45 submissions and the 2019 submissions.

values reported in 2019 ranges between 20% and 32% (with the exception of Nicaragua). Not all of these estimates are fully comparable, however, because there are differences in the sources of error that countries include. This lack of comparability makes it difficult to judge the accuracy of emission estimates based on available uncertainty. Moreover, information on individual error sources would be more useful than aggregate uncertainty estimates in the identification of potential areas for improvement.

The FAO (2018) illustrates the multitude of potential sources of error and how uncertainty analyses often do not include all sources of error, making it difficult to compare FREL/FRL uncertainty estimates. Uncertainties in estimating emissions greatly depend on the error sources considered. Sampling is the most commonly reported source of error, but interpretation (or measurement) errors can be equally significant. Where models are employed, such as allometric equations, root-shoot ratios, soil-carbon stock-change factors and so on, model errors will necessarily affect the estimate (see FAO, 2018a). Not all countries include the same sources of error in their analysis. In fact, an improved assessment of uncertainties is likely to result in a higher aggregate uncertainty because more sources of error are captured in the uncertainty assessment.

2.3.7 Construction approaches and adjustments

Most countries (82%) choose a simple historical average as the construction approach for their FREL/FRLs (Fig. 13). This, however, does not always result in an FREL/FRL value that is equal to the average annual emission/removal over the reference period in the case of delayed emissions or removals that stem

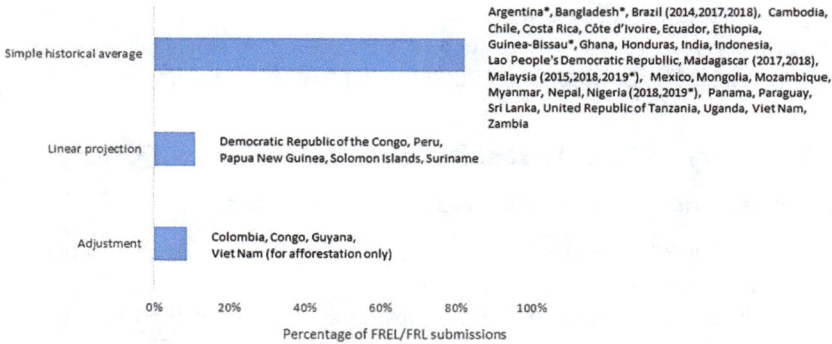

Figure 13 Construction approaches chosen for FREL/FRLs. Note: *Countries with ongoing TAs; construction approach may still change.

from activities during the reference period but continue into the results period, such as the 'inherited' emissions from peatland for Indonesia.

Ghana changed from linear projection to simple historical average as a result of the technical assessment, making this the first case where a country changed its construction approach for that reason.

2.3.8 Consistency of FREL/FRLs with GHG inventories

There are some fundamental differences between GHG inventories and FREL/FRLs. GHG inventories are intended to provide comprehensive estimates of GHG fluxes, while FREL/FRLs are benchmarks for assessing REDD+ performance, often in the context of receiving finance. GHG inventories are entirely based on IPCC guidance and guidelines, and therefore should neither overestimate nor underestimate GHG fluxes. However, the IPCC is not specific about setting a reference level against which to measure results and FREL/FRLs seeking funds may consider the concept of conservativeness.[7] GHG inventory reporting is based on land-use (sub)categories, while FREL/FRLs report by REDD+ activities. Although in some cases IPCC (sub)categories may coincide with REDD+ activities, they do not necessarily match (as discussed in Section 2.3). Generally, GHG inventories aim to be as complete as possible (filling data gaps with Tier 1 estimates), while FREL/FRLs may use a step-wise approach (only including pools and activities for which robust and reliable data are available) or, in the interim, report at the subnational scale. For these reasons, overall net emissions/removals from forests may not be the same in GHG inventories and FREL/FRLs, although this does not necessarily imply that

---

7 When accurate estimates cannot be achieved, the concept of conservativeness suggests that countries should provide estimates that do not overestimate emission reductions or which reduce the risk of overestimation.

they are inconsistent. Since many NDCs take the GHG inventory as a reference or starting point for projections, they may be subject to similar differences, as mentioned above, between the FREL/FRL and GHG inventory.

3 Summary of REDD+ results reported to the UNFCCC

3.1 What's new from REDD+ results submissions

As of early July 2019, the UNFCCC had received 12 REDD+ results submissions from eight countries: this is double the number of countries reporting results since July 2018. Together, the results reported a total of 8.66 billion tCO_2 of emission reductions (ERs) obtained between 2006 and 2017. The large majority of these ERs (94%) are from one country: Brazil (Table 2). Unlike the initial submissions that covered two REDD+ activities only (deforestation and sustainable management of forest), by the end of 2018 all REDD+ activities had been covered in the reported results (yet no single country covers all activities).

The reported net annual ERs consist, on average, of a 32% reduction against the FREL, meaning emissions over the results period are on average 32% lower than emissions in the FREL. The reduction in percentage of the FREL ranges annually between 0% and 69%. Results reported against an FRL consist of a net increase in removals of between 0% and 624%.

Figure 14 shows a few examples of FREL/FRLs and the REDD+ results reported against them.

Figure 15 shows the REDD+ results reported per year by all countries that submitted results to date for the period 2006-2017. These numbers will change as new submissions come in. The annual results reported are highest over the period 2009-2017.

3.2 Results reporting periods and annual variability

Two out of eight countries assess net results over multiple years: Ecuador assesses average emissions over the period 2009-2014, comparing this against the FREL, and Malaysia assesses average removals over the period 2006-2010 without assessing annual variation. All the other countries assess results annually (reporting net results for the combined years). The results periods in the submissions vary between two and seven years but overall, a two-year results period is most common.

As Fig. 17 reveals, most of the annual results assessments show inter-annual variability. This is in line with expectations for the forest sector, as it is influenced by many factors, including annual disturbances (e.g. fires) and climatological differences (e.g. an 'El Niño' year may experience more forest loss than a normal year). For some countries the inter-annual variability can

Table 2 Overview of REDD+ results submitted to the UNFCCC

Year	Submission	Results ('000 tCO$_2$)	Percentage of total results	Average annual results ('000 tCO$_2$)	Results period	Length results period (years)	REDD+ activity
2014	Brazil (Amazon A)	2 971 022	34	594 204	2006–2010	5	Deforestation
2016	Colombia (Amazon I)	28 984*	0.34	14 492	2013–2014	2	Deforestation
2016	Ecuador	28 990	0.34	4832	2009–2014	6	Deforestation
2016	Malaysia	97 470	1.13	19 494	2006–2010	5	Sustainable management of forest
2017	Brazil (Amazon B)	3 154 502	36	630 900	2011–2015	5	Deforestation
2018	Chile	19 362	0.22	1614	2014–2016	3	Deforestation, forest degradation; enhancement; conservation
2018	Colombia (Amazon II)	31 475	0.36	15 737	2015–2016	2	Deforestation
2018	Indonesia	244 892	2.83	16 326	2013–2017	5	Deforestation; forest degradation
2018	Paraguay	26 793	0.31	13 397	2016–2017	2	Deforestation
2019	Brazil (Amazon C)	769 001	9	384 500	2016–2017	2	Deforestation
2019	Brazil (Cerrado)	1 274 723	15	182 103	2011–2017	7	Deforestation
2019	Papua New Guinea	9003	0.10	4502	2014–2015	2	Deforestation, forest degradation; Enhancement**

Notes: *The LULUCF experts 'are of the view that the changes to the national circumstances justifying the adjustment upwards by 10 per cent should not apply to the results reported for 2013–2014' and note that 'the results for 2013–2014 should be considered relative to this conclusion'. The reason is that the condition identified by Colombia to apply the adjustment was the ratification of the peace process prior to the result period.
**Enhancement is included in the scope of the FRL and the results, but for both the activity has been assessed at zero.

Published by Burleigh Dodds Science Publishing Limited, 2021.

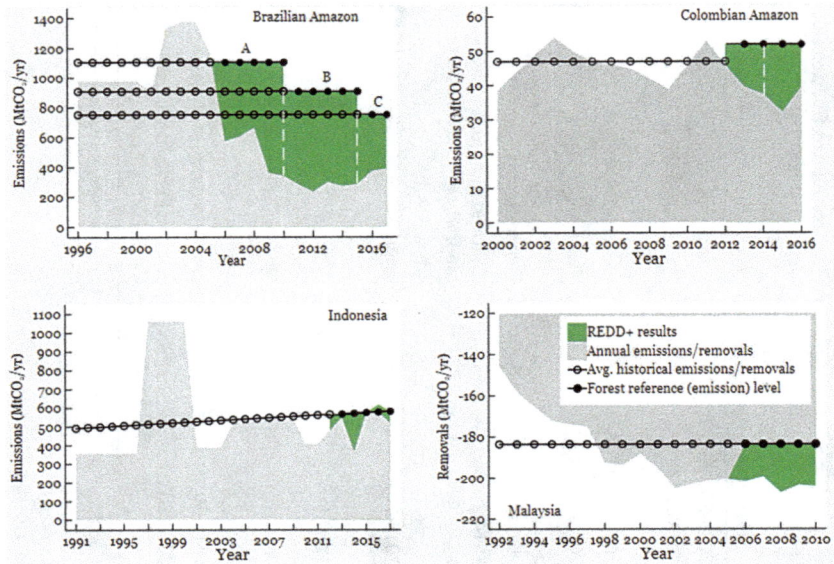

Figure 14 Examples of FREL/FRLs with REDD+ results reported against them.

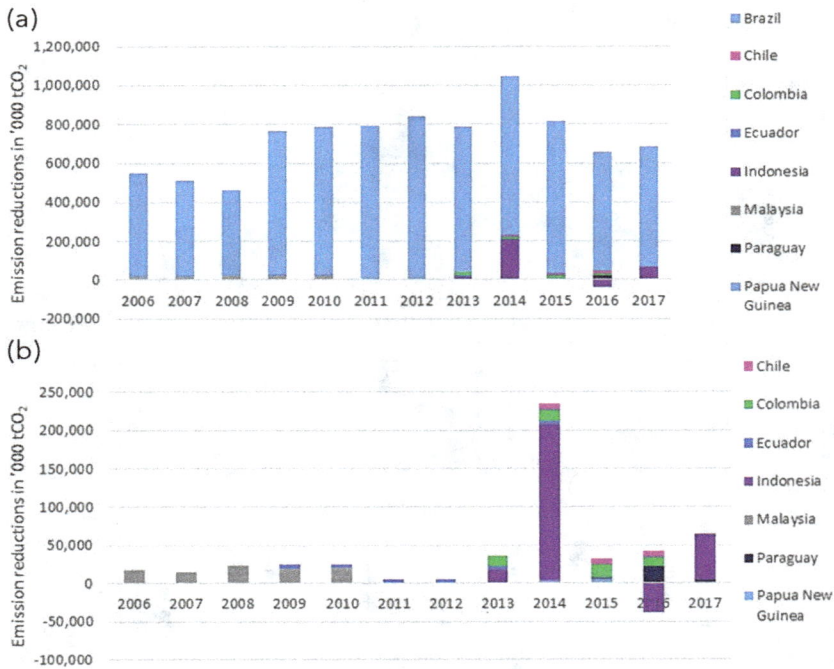

Figure 15 Cumulative REDD+ results reported (a) for all countries, (b) for all countries except Brazil.

Published by Burleigh Dodds Science Publishing Limited, 2021.

Contribution REDD+ activities to emission reductions (without Brazil)

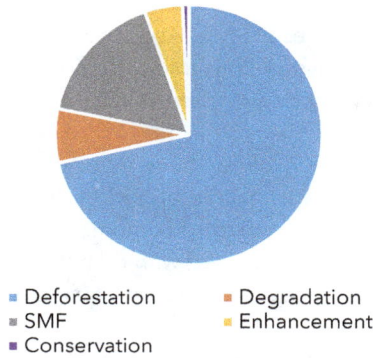

- Deforestation
- SMF
- Conservation
- Degradation
- Enhancement

Figure 16 Contribution of REDD+ activities to total cumulative emission reductions (excluding Brazil). Note: Emissions from peatland are excluded because they are a mix of deforestation and degradation, but not disaggregated by activity.

Percentage of capacity indicators rated as ★ ★ or ★ ★ ★, summed for 16 countries

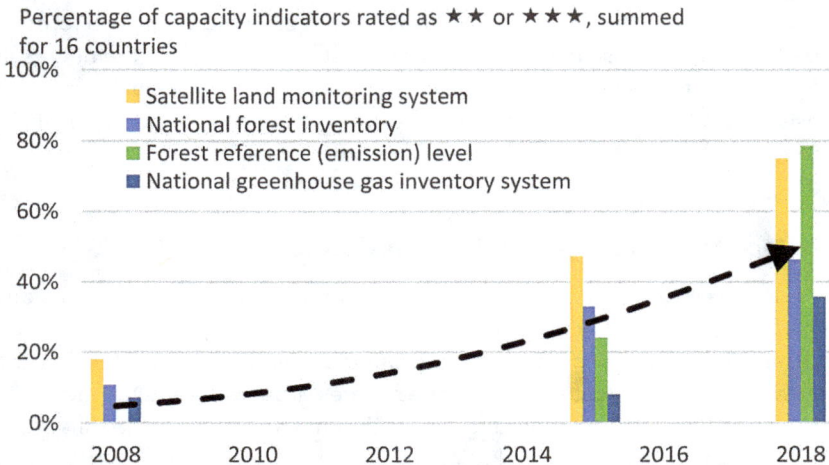

- Satellite land monitoring system
- National forest inventory
- Forest reference (emission) level
- National greenhouse gas inventory system

Figure 17 How capacity development accelerates over time (see Neeff & Piazza 2018, 2019 for a more detailed explanation of the star-rating system).

be relatively large, with years where emissions from activities are actually above the FREL. In total, 182.3 MtCO$_2$ emission increases against the FREL are reported from deforestation and forest degradation. The majority (62%) of these emissions exceeding the FREL come from peat decomposition (which includes both deforestation and forest degradation, but the estimate is not disaggregated by activity), followed by emissions from forest degradation (34%) and deforestation (4%).

3.3 REDD+ activities included for results reporting

Results have been reported for all REDD+ activities (Fig. 16), but the majority of all reported results come from reducing deforestation (99%).

Without considering the results reported by Brazil, the remaining results are still mainly from deforestation (72%), followed by sustainable management of forest (SMF, 16.5%), forest degradation (6.6%), enhancement (4.6%) and conservation (0.8%).

Two countries (Malaysia and Chile)[8] have reported REDD+ results against an FRL expressed in net removals, that is, an FRL for 'plus' activities. For these two examples, the net removals consist of combined emissions and removals. For Malaysia the emissions are associated with timber extraction, and for Chile the emissions stem from degradation inside protected areas, while removals for both originate from growth in (degraded) forest. As such, the results can consist of emission reductions, removal increases or both. For Malaysia's SMF results, 79% are from ERs, stemming from reduced logging following a harvest cap. Chile's net results on forest conservation include a 105% increase in removals from recovering degraded forest in the conservation area and a 185% increase in emissions from forest degradation in the conservation area. In absolute terms, the increase in removals from recovering degraded forest is larger than the increase in emissions from forest degradation, so the net result is positive for this activity. Chile's results on forest carbon stock enhancement consist of 1% from increased removals from the conversion of non-forest land to forest land, and 99% from increased removals in FL-FL.

3.4 Uncertainties around emission reductions

Although the IPCC provides clear guidance on propagating errors for emissions calculation (AD × EF), it does not provide an explicit equation to calculate uncertainties around ERs. Some of the countries that have submitted REDD+ results provide aggregate uncertainty estimates around the REDD+ activities assessed over the results reporting period, but not around ERs (the difference between monitoring and the FREL). The expectation is that uncertainty in estimating ERs will usually be much higher than uncertainty in estimating emissions.

4 Outlook: what's next on MRV for forests?

The UNFCCC Warsaw Framework lays out a continuous improvement process for forest monitoring in countries that participate in REDD+. National forest

---

8 Papua New Guinea also reported REDD+ results against an FRL, but the removals were assessed as zero for both the reference period and the results reporting period.

Published by Burleigh Dodds Science Publishing Limited, 2021.

monitoring systems are meant to start out with current capabilities[9]. The step-wise approach[10] plots a path of continuous improvement, guided by the areas for improvement that UNFCCC submissions and the technical assessment reports identify.

Indeed, progress has been rapid over the last years. The number of countries with REDD+ submissions to the UNFCCC has grown rapidly over the last 5 years (see Section 1.3) and several countries already submitted more than one FREL/FRL to report on technical improvements and new data (see Section 2.1). Furthermore, over the years, technical advancement manifests itself. While pixel counts were more prevalent in early submission, more sophisticated sample-based methods have increasingly come to underlie activity data estimation (see Fig. 4). None of the early submissions reported aggregate uncertainties, but most of the recent submissions did (see Section 2.3). Early submissions often used simplified approaches in establishing emission factors, but more than half of the recent submissions considered post-deforestation biomass (see Section 2.1). Those involved in the UNFCCC technical assessments share a feeling that the technical assessment process has also grown more stringent over the years (see Section 2.2).

In parallel, GHG inventory reporting requirements have become more demanding. While developing countries were initially requested to report only National Communications to the UNFCCC, over time they have been requested to also submit Biennial Update Reports (including GHG inventory updates) that, differently from National Communications, undergo a process of international consultation and analysis. Starting in 2014, countries could submit REDD+ FREL/FRLs, and from 2024 developing countries will be also asked to submit Biennial Transparency Reports.

An assessment of forest monitoring capacity over 2008–2018 has shown learning to accelerate over time (Fig. 17). This observation is consistent with what this chapter concludes about the UNFCCC submissions. With accelerating progress in capacity development and uptake, advancement is initially harder to achieve, but may greatly improve once a foundation has been built.

And further advancement is to be expected. More data is becoming available through an increased selection of available imagery of better quality, more sophisticated algorithms are becoming available to process this data and several open source solutions are becoming available helping countries to access and process this data. A range of new datasets is becoming available that can crucially contribute to countries' forest monitoring efforts. Collecting sample-based data has become much easier with open source solutions like Collect Earth, using Google Earth for visual inspection of satellite imagery. The quality of this interpretation is hugely influenced by the quality of the imagery.

9 Decision 11/CP.19.
10 Decision 12/CP.17, paragraph 10.

So far, analysts often need to rely on medium-resolution imagery, especially from freely available TM Landsat where one pixel is 30 m, which makes it hard to reliably detect change. High-resolution imagery will considerably enhance the robustness of assessments. For example, Planet data with a pixel of, in some cases, as little as three meters were being made available to a range of countries already in late 2019, and wall-to-wall availability of high-resolution satellite imagery is expected during the coming years.

Better data will increasingly become available also on forest biomass densities and the emission factors of deforestation and forest degradation. Over the years, a growing number of countries have conducted national forest inventories. As results become available, countries use these to estimate emission factors. But national forest inventories take long to complete and the next years will see an uptick of more universal application of national forest inventory results.

In a similar vein, new space missions are planned that should deliver data of novel quality and support countries in estimating forest biomass stocks and emission factors. The Global Ecosystem Dynamics Investigation[11] (GEDI) will employ a high-resolution laser ranging from the International Space Station (ISS). Laser ranging will provide information on vegetation height of nearly all tropical and temperate forests. Vegetation height strongly correlates with forest biomass. The availability of GEDI datasets could revolutionize the approach of conducting cumbersome and costly ground-based inventories.

Cloud-processing capabilities are increasingly being leveraged for forest monitoring and collecting activity data. Since FAO launched SEPAL[12] just a few years back, cloud-based computing has supported forest monitoring. Cloud processing is essential because of the need to handle huge datasets required for image acquisition and building country-scale mosaics. This way the hurdle of slow internet and computer processing power can be overcome, which is a key limitation in developing countries. Currently, several teams around the world are working towards enabling multi-temporal processing on the country scale. Sophisticated algorithms crunch numbers across potentially hundreds of satellite images in a dense time series to identify spectral patterns and isolate vegetation changes. Albeit computationally very demanding, dense time-series analysis could revolutionize change detection if applied at a country scale, and cloud-processing platforms are the means to do this.

In sum, current trends in forest monitoring provide grounds for expecting learning and progress to continue. New datasets, and new processing tools are becoming available for both activity data and emission factors. And while the re-submission of FREL/FRLs and results reports is only commencing now, it will provide the basis for continued technical learning as well (see Section

11 https://gedi.umd.edu
12 https://sepal.io

2.1). It will only enhance this trend that increasingly countries also report on results and acquire results-based funding (see Section 3.1). Much stands to be expected from further progress on forest monitoring and reporting for REDD+ during the years to come.

5 References

Birigazzi, L., Gregoire, T. G., Finegold, Y., Cóndor, R. D., Sandker, M., Donegan, E. and Gamarra, J. G. P. 2019. Data quality reporting: good practice for transparent estimates from forest and land cover surveys. *Environmental Science and Policy* 96, 85–94. doi:10.1016/j.envsci.2019.02.009.

FAO. 2013. *National Forest Monitoring Systems: Monitoring and Measurement, Reporting and Verification (M&MRV) in the Context of REDD+ Activities.* UN-REDD programme, Rome. Available at: http://www.fao.org/3/a-bc395e.pdf.

FAO. 2015a. *Technical Considerations for Forest Reference Emission Level and/or Forest Reference Level Construction for REDD+ under the UNFCCC.* UN-REDD programme, Rome. Available at: www.fao.org/3/a-i4847e.pdf.

FAO. 2015b. *Emerging Approaches to Forest Reference Emission Levels and Forest Reference Levels for REDD+.* UN-REDD programme, Rome. Available at: http://www.fao.org/3/a-i4846e.pdf.

FAO. 2015c. *Global Forest Resources Assessment 2015.* Rome. Available at: http://www.fao.org/3/a-i4808e.pdf.

FAO. 2016a. *The Agriculture Sectors in the Intended Nationally Determined Contributions: Analysis.* Environment and Natural Resources Management Working Paper No. 62. Rome. Available at: http://www.fao.org/3/a-i5687e.pdf.

FAO. 2016b. *Map Accuracy Assessment and Area Estimation: A Practical Guide.* National Forest Monitoring Assessment Working Paper No.46/E. Rome. Available at: www.fao.org/3/a-i5601e.pdf.

FAO. 2017. *From Reference Levels to Results Reporting: REDD+ under the UNFCCC.* Forests and Climate Change Working Paper 15. Rome. Available at: www.fao.org/3/a-i7163e.pdf.

FAO. 2018a. *From Reference Levels to Results Reporting: REDD+ under the UNFCCC. 2018 Update.* Forests and Climate Change Working Paper 17. Rome. Available at: http://www.fao.org/3/CA0176EN/ca0176en.pdf.

FAO. 2018b. *Strengthening National Forest Monitoring Systems for REDD+.* National Forest Monitoring and Assessment Working Paper 47. Rome. Available at: http://www.fao.org/3/ca0525en/CA0525EN.pdf.

Foody, G. M. 2010. The impact of imperfect ground reference data on the accuracy of land cover change estimation. *International Journal of Remote Sensing* 30(12), 3275–81. doi:10.1080/01431160902755346.

GCF. 2017. *Terms of Reference for the Pilot Programme for REDD+ Results-Based Payments.* Green Climate Fund. Available at: https://www.greenclimate.fund/documents/20182/1203466/Terms_of_reference_for_the_pilot_programme_for_REDD__results-based_payments.pdf/e26651fc-e216-c8b0-55a1-8eea16a90f39.

GEF. 2018. *Progress Report on the Capacity-Building Initiative for Transparency (CBIT).* Global Environment Facility [Cited 26 April 2019]. Available at: https://www.thegef.org/sites/default/files/documents/EN_GEF.C.55.Inf_.12_CBIT.pdf.

GEF. 2019. *Capacity-Building Initiative for Transparency*. Global Environment Facility [Cited 26 April 2019]. Available at: https://www.thegef.org/topics/capacity-building-initiative-transparency-cbit.

GFOI. 2016. *Integration of Remote-Sensing and Ground-Based Observations for Estimation of Emissions and Removals of Greenhouse Gases in Forests: Methods and Guidance from the Global Forest Observations Initiative*. Edition 2.0. Global Forest Observations Initiative. FAO, Rome.

IPCC. 2006. *2006 IPCC Guidelines for National Greenhouse Gas Inventories*. Prepared by the National Greenhouse Gas Inventories Programme, H.S. Eggleston, L. Buendia, K. Miwa, T. Ngara & K. Tanabe, eds. Vol. 4, Chap. 3.2. Intergovernmental Panel on Climate Change. Institute for Global Environmental Strategies, Kanagawa, Japan.

Lee, D., Skutsch, M. and Sandker, M. 2018a. *Challenges with Measurement and Accounting of the Plus in REDD+*. Report. Available at: http://www.climateandlandusealliance.org/reports/plus-in-redd/.

Lee, D., Llopis, P., Waterworth, R., Roberts, G. and Pearson, T. 2018b. *Approaches to REDD+ Nesting: Lessons Learned from Country Experiences*. World Bank, Washington, DC. Available at: https://openknowledge.worldbank.org/handle/10986/29720.

McRoberts, R. E., Stehman, S. V., Liknes, G. C., Næsset, E., Sannier, C. and Walters, B. F. 2018. The effects of imperfect reference data on remote sensing-assisted estimators of land cover class proportions. *ISPRS Journal of Photogrammetry and Remote Sensing* 142, 292–300. doi:10.1016/j.isprsjprs.2018.06.002.

Neeff, T. and Piazza, M. 2018. *Ten Years of Capacity Development on National Forest Monitoring for REDD+ – Much Achieved Yet More To Do*. Rome. Available at: http://www.fao.org/3/CA1741EN/ca1741en.pdf.

Neeff, T. and Piazza, M. 2019. Developing forest monitoring capacity – progress achieved and gaps remaining after ten years. *Forest Policy and Economics* 101, 88–95. doi:10.1016/j.forpol.2018.10.013.

Olofsson, P., Foody, G. M., Stehman, S. V. and Woodcock, C. E. 2013. Making better use of accuracy data in land change studies: estimating accuracy and area and quantifying uncertainty using stratified estimation. *Remote Sensing of Environment* 129, 122–31. doi:10.1016/j.rse.2012.10.031.

Olofsson, P., Foody, G. M., Herold, M., Stehman, S. V., Woodcock, C. E. and Wulder, M. A. 2014. Good practices for estimating area and assessing accuracy of land change. *Remote Sensing of Environment* 148, 42–57. doi:10.1016/j.rse.2014.02.015.

Pagliarella, M. C., Corona, P. and Fattorini, L. 2018. Spatially-balanced sampling versus unbalanced stratified sampling for assessing forest change: evidences in favour of spatial balance. *Environmental and Ecological Statistics* 25(1), 111–23. doi:10.1007/s10651-017-0378-y.

Pearson, T. R. H., Brown, S. and Casarim, F. M. 2014. Carbon emissions from tropical forest degradation caused by logging. *Environmental Research Letters* 9(3), 034017. doi:10.1088/1748-9326/9/3/034017.

UNFCCC. 2013. *Warsaw Framework for REDD+*. United Nations Framework Convention on Climate Change. Available at: https://unfccc.int/topics/land-use/resources/warsaw-framework-for-redd-plus.

UNFCCC. 2019a. *REDD+ Safeguards*. United Nations Framework Convention on Climate Change. Available at: https://redd.unfccc.int/fact-sheets/safeguards.html.

UNFCCC. 2019b. *What is Transparency and Reporting?* United Nations Framework Convention on Climate Change. Available at: https://unfccc.int/process-and-meet ings/transparency-and-reporting/the-big-picture/what-is-transparency-and-reporting.

World Bank. 2008. *Pilot Program for Climate Resilience (PPCR).* Washington, DC. Available at: https://climatefundsupdate.org/pilot-program-for-climate-resilience/.

World Bank. 2013. *World Development Indicators 2013.* Washington, DC. Available at: http://documents.worldbank.org/curated/en/449421468331173478/pdf/76824 0PUB0EPI00IC00PUB0DATE04012013.pdf.

World Bank. 2013/2017. *Operational Guidance for Monitoring and Evaluation (M&E) in Climate and Disaster Resilience-Building Operations.* Washington, DC. Available at: http://documents.worldbank.org/curated/en/692091513937457908/pdf/122226 -ReME-Operational-Guidance-Note-External-FINAL.pdf.

Chapter 4

Advances in understanding the role of forests in the carbon cycle

Matthew J. McGrath and Anne Sofie Lansø, Laboratoire des sciences du climat et de l'environnement, France; Guillaume Marie, Vrije Universiteit Amsterdam, The Netherlands; Yi-Ying Chen, Academia Sinica, Taiwan; Tuomo Kalliokoski, University of Helsinki, Finland; Sebastiaan Luyssaert and Kim Naudts, Vrije Universiteit Amsterdam, The Netherlands; Philippe Peylin, Laboratoire des sciences du climat et de l'environnement, France; and Aude Valade, Ecological and Forestry Applications Research Centre, Spain

1 Introduction

Forests cover around 31% of the global land area (Food and Agriculture Organization of the United Nations, 2015), and human societies have evolved to depend on them for a variety of ecosystem services, such as air and water quality regulation, wood production, and recreation. Over the past half century, the recognition that increased levels of atmospheric carbon dioxide are warming the global climate has led to forests being viewed as potential natural ways to reduce carbon dioxide concentrations in the atmosphere. Forests' complex interactions with local and global climates, however, make predicting the impacts of changes in forest cover and composition challenging. Managing forests for carbon storage leads to the risk of losing other benefits, in extreme cases even failing to achieve climate objectives due to biogeophysical

http://dx.doi.org/10.19103/AS.2019.0057.06

effects, which can reduce or eliminate cooling from removal of atmospheric carbon dioxide. In such a context, the science behind carbon storage in forests develops even greater importance.

2 The importance of forest carbon content

Advances in our understanding of how and where carbon is stored in forest ecosystems and how it impacts climate fuel international policies and agreements, which in turn leads to higher demand for scientific knowledge of this system. In this way, scientific priorities and governmental policies reflect the interactions between forests and the atmosphere: forests change climate and climate changes forests. Forests also provide additional ecosystem services on which society depends, and these should not be lost in the push to store carbon. The link between forests and climate is why society depends on understanding how carbon moves through forest ecosystems. Our understanding of how forests impact local, regional, and global climate comes from a combination of observations, measurements, and computer models, and involves complicated feedbacks between biogeochemical and biogeophysical effects (cf., Pielke Sr., et al., 2002; Bonan, 2008). These issues of competing effects and human activities are introduced here, but will be revisited throughout the rest of this chapter as reoccurring themes.

2.1 Interactions between forests and climate

Human activities play a crucial role in the forest-climate interaction, most notably by changing land cover (deforestation, afforestation, reforestation) or modifying forest properties, although the net climatic effect of adding and removing forests remains unclear. Forests are generally regarded as a net carbon sink (explored in more detail in later sections), and widely viewed to lead to cool the climate by removing and storing carbon dioxide from the atmosphere: IPCC (2018) note the potential for afforestation (a form of land-use change) to remove up to 3.6 $GtCO_2$ year^{-1} , while Le Quéré et al. (2018) report that general land-use change (primarily deforestation) was responsible for around 1.5 Gt of carbon emitted into the atmosphere each year over the period 2008-17. In addition, many authors have demonstrated significant climate effects of deforestation due to biophysical effects, including Alkama and Cescatti (2016), who used observational data from areas undergoing recent land-cover transitions; Bright et al. (2017), who used FLUXNET site data to develop a global model of temperature redistribution; and Bala et al. (2007), who used global climate simulations to demonstrate that global-scale deforestation would have a net cooling effect on the climate, due to biophysical effects overwhelming warming effects from changes in the carbon cycle. In temperate regions, Bonan (2008) noted that the competition between evapotranspiration and snow masking leads to uncertain net impacts of deforestation. In addition, many early studies did not include the impacts of past land-use decisions, which may underestimate the role of the carbon cycle (Pongratz et al., 2011), and both Luyssaert et al. (2014) and Erb et al. (2017) demonstrate that wood harvest and species selection result in significant changes to local climate on the same magnitude as complete conversion of forest to other land-cover types. Therefore, while it is well-agreed that forests impact climate (Bennett and Barton, 2018), specific results are sensitive to the exact situation of the forest in question.

To further complicate the situation, forests can form both positive and negative feedback loops with the climate. Positive feedback loops happen when warming increasing carbon dioxide production in forests, which raise global carbon dioxide levels and cause more warming. Melillo et al. (2017) demonstrated that increasing the soil temperature by five degrees in a mid-latitude hardwood forest increased fluxes of carbon out of the soil. When extrapolated to a global scale, warming-induced emission of soil carbon dioxide over the twenty-first century would match fossil fuel emissions over the past two decades (Melillo et al., 2017). Lindner et al. (2010) synthesized existing literature covering Europe and found that future climate is expected to result in increasing forest production in some regions while other regions would suffer due to increased risk of drought and disturbances. For boreal forests, the net effect was to increase production, though globally this was negated by effects in tropical forests (Bonan, 2008). Negative feedback loops could arise from biogenic volatile organic compounds (BVOCs), produced by boreal forests in large quantities (Tunved et al., 2006), and which are known to lead to cloud formation and thus modify the atmospheric radiation budget (Gordon et al., 2017); to our knowledge, direct studies using climate models with explicit representation of these processes have not yet been carried out (Arneth et al., 2010).

2.2 Policies regarding carbon storage in forests

Due to the importance of carbon to climate, forests as tools to mitigate climate change through carbon storage have received attention at the highest international level. Under Article 2 of the Kyoto Protocol, Parties agreed, in 1998, to implement and elaborate forestry management practices, including afforestation and reforestation, to protect and enhance sinks and reservoirs of greenhouse gases (Member countries of the United Nations, 1998). Article 5 of the Paris Agreement explicitly recognizes the role of forests as greenhouse gas sinks and reservoirs (Member countries of the United Nations, 2015). Article 4 of the Paris Agreement instructs Parties to prepare nationally determined contributions (NDC) they will use to achieve the long-term temperature goal set out in Article 2. Out of the ten countries with the largest boreal and temperate forest area (Russian Federation, Canada, the United States, China, Sweden, Finland, France, Chile, Norway, and New Zealand (Food and Agriculture Organization of the United Nations, 2015)), four explicitly mention forests in their NDC, while five imply forests will be used to meet objectives (note that the European Union submitted a collective NDC); only Russia failed to submit an NDC (UNFCCC, n.d.). In Europe, European Academies Science Advisory Council (2017) noted that 'Maintenance and appropriate enhancement of forest resources and their contribution to global carbon cycles' was one of six criteria of sustainable forest management adopted by the 46 Member States to the 7th Forest Europe Ministerial Conference in 2015. International organizations thus clearly see a role for forest in climate change mitigation across boreal and temperate zones.

The IPCC routinely acknowledges the role of forestry (primarily afforestation and reforestation) in carbon sequestration (IPCC, 2014, 2018). Large-scale afforestation programs are not without critics, however. While afforestation and reforestation can remove carbon dioxide from the atmosphere, European Academies Science Advisory Council (2018) cautioned that technology does not currently have the potential to replace emission reductions at the scale required for effective mitigation of climate change. Through the use of state-of-the-art land-surface computer models coupled to an atmospheric general circulation model, Luyssaert et al. (2018) showed that realistic implementation of forest management decisions in Europe (changing management method and/or tree species) may achieve reduction of atmospheric carbon dioxide concentrations without meeting the remaining climate-oriented objectives of the Paris Agreement, although BVOC production and impacts on cloud formation were not included. The term 'best available science' appears in the Paris Agreement four times (Member countries of the United Nations, 2015), highlighting the importance of science to this discussion, and more scientific evidence is needed to ensure that policy decisions achieve their desired effects.

3 Monitoring forest carbon

Forest managers, policy makers, and scientists refer to carbon in forests in two distinct ways: (1) carbon storage in forests, as a tool to mitigating climate change, and (2) carbon emissions from forests, as something to avoid and to prevent accelerating climate change. Discussions about carbon storage generally take a long-term view, stretching out to decades and centuries. Discussions around carbon emissions can likewise encompass long time horizons (in terms of decomposing biomass on the forest floor or emission from soils), but they can also focus on shorter time scales, as when a forest is clear-cut for conversion to agricultural lands. This section considers how scientists estimate both carbon

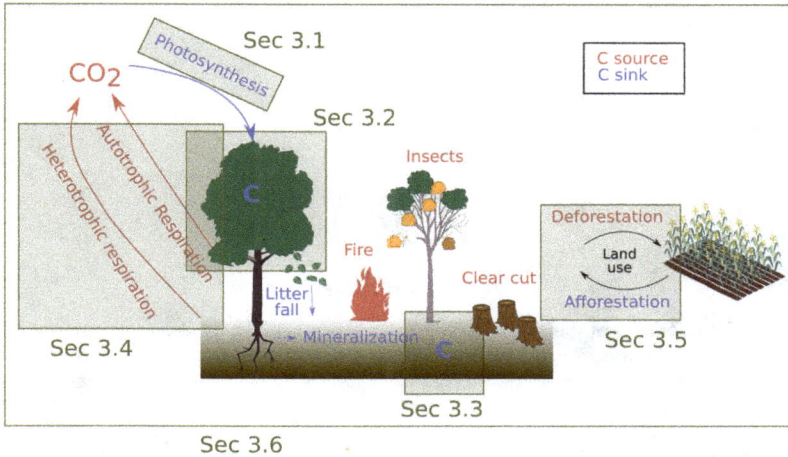

Figure 1 The primary carbon fluxes and stores considered in Section 3. Source: modified from Bonan (2008). Reprinted with permission from AAAS.

storage and emissions. This section partitions the carbon cycle in forests into six tasks that forest managers and scientists typically follow, though researchers break the cycle down in various ways depending on the questions they are exploring (cf. Bellassen and Luyssaert, 2014; European Academies Science Advisory Council, 2017). The next paragraph explores terminology and how it applies to our breakdown of the carbon cycle.

We distinguish between two key components of the carbon cycle: fluxes and stocks (Fig. 1). Stocks are typically straightforward to understand (e.g., the amount of carbon stored in trees and soils, covered in Sections 3.2 and 3.3, respectively). Fluxes describe the flow of carbon between storage pools. We follow the flux definitions laid out by Chapin III et al. (2006), a standard reference in the field. For space reasons, we do not reproduce their definitions here.

The priority of this section is to provide a sampling of methods, with enough descriptive details for a few of them to give the reader a solid introduction for estimating carbon in forests and forest landscapes. Many of the methods (in particular the models) can be used for multiple tasks listed below. This list should not be considered exhaustive in any way, and certain classes of models and measurements are completely sacrificed to provide illustrative examples of others. A broad comparison of the spatial and temporal scales of the approaches is given in Fig. 2.

3.1 Estimating carbon assimilation from the atmosphere

Trees store carbon by first removing it from the atmosphere via photosynthesis, referred to as gross primary production (GPP). Two primary methods exist for

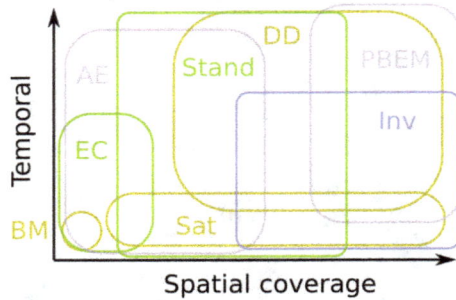

Figure 2 An overview of the spatial and temporal applicability of the methods reviewed here. Spatial coverage ranges from individual trees to the global land surface, while temporal scales range from single moments to hundreds of years. BM: biomass sample; EC: eddy-covariance; AE: allometric equations; Stand: stand-level models; DD: data-driven models; Sat: satellite data; PBEM: process-based ecosystem models; Inv: atmospheric inversions. Boundaries are approximate. Colors are only used to distinguish methods.

estimating GPP in the field: the eddy-covariance (EC) micro-meteorological technique and ecology-based biometric methods (BM). Campioli et al. (2017) carried out a comparison between the two methods, showing that they result in comparable estimates for forest GPP across the globe. EC measures the net flux of carbon dioxide from the canopy to the atmosphere (Baldocchi, 2003). BM, on the other hand, require much more intensive labor and track the carbon flows across the forest by measuring carbon pools and increments, for example litter production, losses of organic carbon, and respiratory carbon losses (Hamilton et al., 2002).

Due to challenges of establishing extensive field monitoring networks in forests, researchers have been developing ways to estimate forest productivity from satellite platforms and from global ecosystem models. Yang et al. (2015) showed that solar-induced fluorescence (SIF) correlates with very high resolution ground-based measurements of GPP in a deciduous temperate forest. Using relationships between physical environmental variables and plant-specific parameters (e.g. Farquhar et al., 1980), modelers can predict the GPP of forests at regional and global scales (e.g. Jung et al., 2007). Recent efforts by MacBean et al. (2018) incorporate SIF data as a constraint on the GPP of ecosystem models, tying together methods for estimating forest GPP where no ground measurements are available.

3.2 Estimating forest biomass

Direct sampling of biomass is arguably the most accurate method for small areas. Whole harvested trees can be oven-dried and weighed (Bond-Lamberty et al., 2002). Fine roots can be sampled by taking soil cores (e.g. Ostonen et al., 2017; Solly et al., 2018), while coarse roots can be sampled by ground-penetrating radar (Molon et al., 2017), along with root excavation and soil-pits (Addo-Danso et al., 2016). Maaroufi et al. (2016) arranged litter tray traps, positioned two centimeters off the forest floor and emptied once a month, to catch falling debris. In addition, several areas were cleared of branches to estimate the input of dead branches. Notable disadvantages of these methods include the inability to scale to larger areas due to the labor involved, and the destruction of the samples (e.g. their removal from the ecosystem).

From direct measurements of trees, one can derive allometric equations that relate the amount of biomass in a tree to an easily measurable quantity like the diameter of the trunk. Such equations form the foundation of field biomass estimates (e.g. McMahon et al., 2010; Melillo et al., 2011; Pretzsch et al., 2014; Ford and Keeton, 2017; Urbano and Keeton, 2017; Shao et al., 2017; Gordon et al., 2018; Vieilledent et al., 2018; Yuan et al., 2018). From equations established at the genus or species level for a geographical region, researchers can estimate tree size by measuring the increase of stem width over a year, or by counting the number of trees per given area; both of these are significantly less labor-intensive than directly measuring tree biomass, and do not require killing the tree. In addition, sampling data for tree diameter and species exists in immense quantities through systematic forest inventories. Sets of equations have been developed for species across the boreal and temperate zone, including the United States (Jenkins et al., 2003), Australia/New Zealand (Beets et al., 2012; Sillett et al., 2015), China (Wang, 2006), and Canada (Ung et al., 2008). From these, one can estimate the amount of living aboveground biomass (or dead biomass, using similar methods and an estimated degree of decomposition, for example, Gordon et al., 2018) in a stand and convert that to carbon (McGroddy et al. (2004) assumed carbon to be 47% of the biomass, though 50% is more commonly used). Sullivan et al. (2017) cautioned that, despite the power of allometric relationships, they do not allow for variations in tree architecture with forest structure.

Satellite and airplane platforms open up possibilities to map the forest biomass of large areas (Goetz et al., 2009; Zhang et al., 2014; Coomes et al., 2018; Sadeghi et al., 2018; Vastaranta et al., 2018). Sadeghi et al. (2018) estimated the biomass of a Canadian boreal forest covering an area on the order of 100 square kilometers using satellite-based spectrometry and interferometric synthetic aperture radar (InSAR) with spatial and temporal overlap, combined with field plot measurements. The spectrometry product Landsat records how the land cover absorbs various wavelengths of light, giving an indication of the amount of vegetation present. The InSAR provided a digital elevation model that gave a reasonable prediction of the difference between the top of the canopy and the ground surface. The authors developed a statistical model to predict biomass at 20 meters resolution using combined variables from both satellite products. Such efforts allow for improved biomass prediction when scaling up plot-level measurements to whole forests and filling in gaps for areas with no measurements available. Light Detection and Ranging (LiDAR) provides another form of light-based measurement with the capacity to revolutionize biomass estimation in forests, and is covered in more detail in Section 7.2.

While the number of applicable computer modeling approaches to predict forest biomass is enormous (cf. Barredo et al. (2012), Charru et al. (2017), and Schelhaas et al. (2018) for data-driven models, or more complex models like CBM-CFS3 (Kull et al., 2016); CO2FIX (Schelhaas et al., 2004); ForClim (Gutiérrez et al., 2016); EFISCEN (Schelhaas et al., 2007; Verkerk et al., 2016); the process-based 4C model (Lasch-Born et al., 2015)), we focus on more generalized process-based ecosystem models which have a larger number of explicit ecological and physical mechanisms (e.g. ORCHIDEE-CAN (Naudts et al., 2015); PnET-CN (Peters et al., 2013), JSBACH (applied to Finnish forests in Peltoniemi et al. (2015)); JULES (Harper et al., 2016), CLM (Oleson et al., 2010)). The line between 'process-based ecosystem models,' 'dynamic global vegetation models,' and 'land surface models with dynamic vegetation' is not always clear. To take one example, ORCHIDEE-CAN uses climate data based on the latitude and longitude of the pixel to compute the amount of carbon assimilated by groups of identical trees of different diameter classes at 30-minute intervals. These trees can be either species or more generic plant functional types (e.g. temperate conifers). At the end of every day, this carbon is allocated to different pools representing various parts of the tree (roots, sapwood, and leaves), and the various pools undergo turnover to convert the carbon into other pools (such as hardwood and litter). Once a year, the grid square can undergo management, wherein some quantity of woody biomass is removed from site to simulate thinning or full clear-cuts. The model thus stores the amount of carbon available in each of the pools at daily resolution. While simplified, ecosystem models share many of these concepts, depending on the specific model. Some ecosystem models can be coupled directly to atmospheric circulation models, providing insight into forest growth at large scales and the resulting climatic impact; such models are often called 'land-surface models' and generally contain less complexity (in terms of processes included and their descriptions) than stand-level ecosystem models.

3.3 Estimating carbon storage in soils

Carbon enters the soil through decomposing litter and roots. Like the biomass pools, carbon storage in soils can be estimated through measurements and computer models. Some of the methods are quite labor-intensive (cf., Strickland et al., 2010; Karst et al., 2017; Shabaga et al., 2017; Stendahl et al., 2017). Karst et al. (2017), for example, measured the amount of carbon that flows from the roots to the soil in aspen forests by planting seedlings in pots, flushing the roots with distilled water after photosynthesis was measured, and then collecting and analyzing the runoff to determine the carbon content. Shabaga et al. (2017) determined soil carbon content by extracting 20 cm deep columns in a hardwood forest in Central Ontario (Canada). Stendahl et al. (2017) measured the amount of carbon in the soil of boreal forests through a relationship between the carbon concentration of the soil, bulk density, soil layer thickness, and the volume percentage of stones and boulders.

Using computer models to estimate carbon soil storage saves vast amounts of labor and permits extension to remote areas, though such models require extensive validation with measurements and experiments. The models have different levels of complexity and scale (e.g. the landscape, a stand, or a tree), with trade-offs being made between detail and computational cost depending on the research question. For example, the Yasso07/Yasso15 models (Tuomi et al., 2009; Ziche et al., 2019) require litter input data to the soil across Europe and follow the chemical decomposition of these inputs based on measurements under different climatic conditions. Gonsamo et al. (2017) apply the Integrated Terrestrial Ecosystem Carbon Cycle model (InTEC) (Chen et al., 2000) to unmanaged forests in northern Ontario, including multiple soil carbon pools, nitrogen decomposition, and drainage in addition to aboveground photosynthesis. Many of the process-based ecosystem models in the previous section include information on soil carbon, as well.

Estimates of soil organic carbon in forests are very imprecise, mostly due to variability. Loss of carbon through various decomposition processes following disturbances (deforestation, harvesting) is rapid but sequestration in soil occurs slowly (hysteresis effect), so this is more of an issue of estimating stocks than fluxes. Three areas where better estimates are needed are (1) deep carbon in soils such as Spodosols (podzols) (cf. Stone et al., 1993); (2) forested wetlands (Trettin and Jurgensen, 2003); and (3) carbon buried in sediments (cf. Breithaupt et al., 2012).

3.4 Estimating carbon loss from forests

Forests lose carbon through a variety of processes, including autotrophic respiration (from the trees themselves), heterotrophic respiration (from herbivores or microbes), and leaching from groundwater or surface water, as well as decomposition following disturbances both natural (pests, storms and fire) and human-made (harvest, thinning, conversion to agriculture). Forest disturbances are covered in more detail in Section 4.

To quantify carbon losses through respiration, Schindlbacher et al. (2015b) measured the carbon dioxide concentration above extracted soil cores contained in laboratory measurement chambers from mountain forests in Austria, warming both the cores and the forest soil to examine possible impacts of climate warming. In the field, soil respiration can be measured by trapping and measuring the CO_2 concentration of the air above the soil with an infrared gas analyzer (e.g., Shabaga et al., 2017; Schindlbacher et al., 2015a). Bathiany et al. (2010) used a computer model that includes two soil pools, calculating soil respiration according to a parameter that relates the increase caused by a 10 degree increase in temperature (Q10, c.f. Meyer et al. (2018)) and a linear dependence on soil moisture. Naipal et al. (2018) used a process-based global ecosystem model to look at lateral carbon transfer due to soil erosion. These methods are some ways to estimate non-disturbance-based carbon losses in temperate and boreal forests.

3.5 Estimating change in forest area

Carbon management crosses ecosystem boundaries. The biggest examples of this are land-use, land-use change and forestry (LULUCF) activities, where conversion of a forest to agricultural land results in a net emission from a forest, despite that the land is no longer forest. In order to estimate this, models rely on maps indicating forest areas, so-called 'land cover maps.' The product Globcover (Arino et al., 2008) takes in spectral data from the MERIS satellite instrument to compile a map of 23 land-cover types. Future land-cover maps rely on models incorporating biophysical and socioeconomic conditions (Prestele et al., 2016). Landsat data has been used to produce a high-resolution map showing global tree cover (Hansen et al., 2013), though the definition of forest as 'all vegetation taller than 5m in height' includes tree plantations which do not exhibit the benefits of forests (Tropek et al., 2014). Merging plot-level data from international forestry databases and large international inventories along with a model to predict densities across the globe, Crowther et al. (2015) produced a map showing global tree density. A map showing the spatial distribution of tree species over Europe has been produced by statistically combining ICP-Forest Level-I plot data and National Forest Inventory plot data (Brus et al., 2012).

From these maps and others, researchers can identify areas of forest loss. The cause of a gap appearing in a forest canopy from one year to the next is not evident. Distinguishing between gaps caused by harvesting and regrowth or land-use change such as conversion to agriculture requires a multi-year comparison by pixel, appropriately rectified and adjusting for spectral differences between scenes. Hansen et al. (2013) did this for their own forest map, identifying pixels where the tree canopy had been completely removed or had completely appeared when comparing the same pixel in different years. The accuracy of gross rates of forest loss and gain from remote-sensing-based maps can be estimated with sound statistical practices (Olofsson et al., 2014). Hot-spots where forest loss is more likely to be driven by underlying spatial process (i.e. not random events) have been identified through advanced statistical analysis techniques applied to existing forest loss maps (Harris et al., 2017). Though the authors only applied it to tropical areas, the technique has broader applications, including temperate and boreal forests. Such efforts help determine country-level changes in forest cover which can be used to compute how much carbon is lost or gained through LULUCF activities.

3.6 Estimating net carbon flux from forests

This subsection focuses on the estimation of carbon fluxes from large swathes of forest instead of plot-level measurements, identifying methods that researchers use to account for boreal and temperate forest carbon fluxes from sectors, countries, and the globe. It is important to distinguish between published estimates in the scientific literature of forest sinks and sources, and 'reporting' and 'accounting' carried out by countries under international climate agreements. Estimates in the literature may include both natural and anthropogenic fluxes, while GHG reporting includes only anthropogenic fluxes (Grassi et al., 2018). In addition, accounting refers to 'the comparison of emissions and removals with the target and quantifies progress toward the target' (Grassi et al., 2018). This distinction can be difficult to make in practice, as in the case of human-induced atmospheric changes resulting in increased carbon storage in natural forests. Atmospheric inversion models (cf. Gerbig et al., 2003; Gurney et al., 2004; Matross et al., 2006; Chevallier et al., 2010; Ciais et al., 2010; Uglietti et al., 2011; Saeki et al., 2013; Thompson and Stohl, 2014; Steinkamp et al., 2017), while providing a necessary constraint on so-called 'bottom-up' approaches, are excluded here as they do not specifically deal with forests. Many of the models described in previous subsections can also be used for these purposes.

One way that researchers estimate net carbon fluxes from forests are through bookkeeping models, widely used to account for emissions from land-use, land-use change, and forestry (LU LUCF), where carbon leaves forests due to human activities (c.f., Houghton et al., 1983; Reick et al., 2010; Gasser and Ciais, 2013; BLUE (Hansis et al., 2015)). The BLUE model is spatially and temporally explicit, at a resolution determined by the input data (currently typically a spatial resolution of 0.25 degrees and a timestep of one year). The model tracks carbon in pools labeled by cover type (e.g. primary forest, grassland), transition type (e.g. harvest, abandonment), pool type (e.g. slow soil, biomass), and PFT through the use of equilibrium estimates of carbon densities derived from observations or vegetation models. The values of the pools change through prescribed growth and decomposition curves. When the driving maps indicate a transition happens, BLUE shifts carbon between the pools as a function of the transition and the values of the pools change through prescribed growth and decomposition curves that strive toward reestablishing the equilibrium. Through this, BLUE can track emissions of carbon from LULUCF activities, thus isolating fluxes due to direct human interference from carbon fluxes caused by changes in environmental conditions.

4 Mechanisms driving forest carbon storage

Understanding the underlying mechanisms driving forest carbon sequestration allows us to better assess the future capability of the global forest to store carbon. Ecological and physiological drivers are inherently linked to disturbances of the forest biome and their impacts range from low to high. It has been estimated that 44% of the world's forest carbon stock resides in the soil, 42% in live biomass (above and below ground), 8% in deadwood, and 5% in litter, although these fractions vary considerably between geographical regions (e.g. tropical and boreal forests store carbon in different places, with boreal forests having much larger soil stocks) (Pan et al., 2011a). In boreal and temperate forests the carbon cycle is driven by forest management and afforestation, with evidence from forest plots, yield tables and biogeochemical models indicating that climate change, CO_2 fertilization and N deposition notably impact long-term carbon storage (Pan et al., 2011a; Pretzsch et al., 2014). How much each contributes, however, is still under debate (e.g. Magnani et al., 2007; Keenan et al., 2013; Campioli et al., 2015).

4.1 Ecological/physiological/environmental drivers

By definition, a tree can only grow if more carbon is stored than released. Forest carbon storage starts from the leaf's capacity to capture CO_2 from the atmosphere. From this available carbon, trees create complex molecules to build biomass. In addition, trees respire, releasing carbon to grow and maintain living tissue. In turn, dead tissue from trees falls to the ground, where carbon can either be respired or mineralized by bacteria that progressively increase the amount of carbon sequestered in the soil. The age of a forest is inherently linked to the time since last disturbance, and can thus serve as proxy for the net carbon accumulated by the forest ecosystem (Pan et al., 2011b).

Young forests are more productive than older forests (Ryan et al., 1997), having an accelerating assimilation rate in their early successional development that peaks when canopy closure is reached (Pan et al., 2011b). Despite this high productivity, young even-aged stands do not store carbon in their first years (less than ten years, though this can be longer in the boreal regions). The net ecosystem productivity (NEP) is negative during this time with CO_2 being released to the atmosphere (Pregitzer and Euskirchen, 2004) due to high heterotrophic respiration. For even-aged forests the heterotrophic respiration is typically high in the early successional stage because of site preparation leaving litter and woody debris on the forest floor to be respired (Luyssaert et al., 2008), though stand-clearing disturbances like fire or harvest result in differences in carbon dynamics due to the amount and state of the leftover material.

NEP peaks during the slower-growing middle successional stage (Fig. 3), before declining during the mature stage. Although similar NEP patterns are seen, observation and theory do not imply that age-related NEP trajectories are the same for all biomes and species (Gough et al., 2016). The timing and magnitude of the NEP peak and the subsequent decline varies between forest ecosystem (Curtis and Gough, 2018) depending on geographic location, prevalent disturbance, and dominant tree composition (Curtis and Gough, 2018). NEP of temperate evergreens increases more rapidly in the early stages than deciduous, and reaches peak NEP values around year 10–30 (Pregitzer

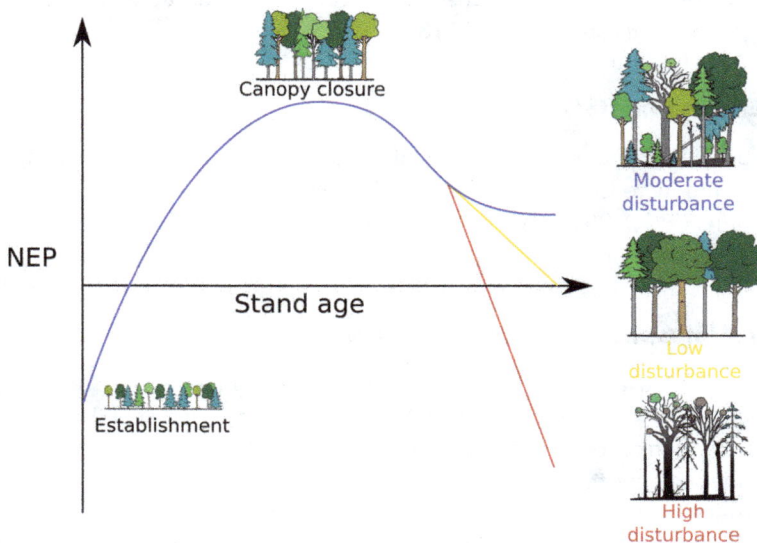

Figure 3 Net ecosystem productivity (NEP) as a function of stand age following disturbance. Note the low (even negative) NEP for young stands due to respiration from decomposition of vegetation left on site following the disturbance (natural or anthropogenic). Note as well the possibility for higher NEP in old, moderately disturbed stands compared to old stands with low disturbance, a counterintuitive result, due to more complex and physiologically efficient canopies resulting from moderate disturbances (Hardiman et al., 2013; Curtis and Gough, 2018), although NEP can exhibit wide variability in older boreal forests (Taylor et al., 2014).

and Euskirchen, 2004), while deciduous forests do not peak until around 30 years of age (Gough et al., 2016). Less favorable climatic conditions in boreal regions result in peak NEP of boreal evergreen forests only occurring in years 71–120 (Pregitzer and Euskirchen, 2004). The declining age-trend of NEP is greater for evergreen temperate than deciduous, and for some deciduous stands only a slight decrease in NEP is observed (Pregitzer and Euskirchen, 2004; Luyssaert et al., 2008; He et al., 2012; Gough et al., 2016; Curtis and Gough, 2018).

During middle and mature stages, the fast-growing early successional species are overtaken by slower-growing species, and the death of the fast-growing species increases the heterotrophic respiration as litter fuels carbon losses from decomposition within the forest biomes (Gough et al., 2016). This adds complexity to the canopy structure and forest ecosystem, which can be further amplified by low and moderate impact disturbances. Structural complexity is hypothesized to increase resource efficiency, which can sustain high productivity in later stages of succession (Hardiman et al., 2013; Curtis and Gough, 2018). Synthesis-based observational studies point toward old-age forests remaining sinks and continuing to add additional carbon to their large existing stores (Pregitzer and Euskirchen, 2004; Luyssaert et al., 2008; Gough et al., 2016; Curtis and Gough, 2018).

4.2 Environmental drivers

During the last century, environmental ecosystem drivers have changed, most notably the increase in atmospheric CO_2 concentrations and meteorological conditions (IPCC, 2014). Nitrogen deposition and soil disturbance also have seen notable changes.

4.2.1 Atmospheric CO_2 concentrations

Free Air CO_2 Experiments (FACE) have shown that higher carbon accumulation can be sustained under elevated CO_2 levels with increases of NPP from 20 to 50% in temperate forest ecosystems (Ciais et al. (2013) and references therein). The variability and even the lack of response at some sites can be ascribed to nutrient limitations, climatic conditions, and species composition (Norby et al., 2010). However, the short time span of FACE experiments only allows for single-generation assessment, and does not answer questions related to plant plasticity and adaptation across multiple generations. With a meta-analysis using vegetation grown at CO_2 springs, Saban et al. (2019) have recently confirmed that the long-term tree responses to increased levels of atmospheric CO_2 are consistent with the conclusions drawn from the FACE experiments with regards to increases in photosynthesis and leaf starch, and decreases of stomatal conductance, leaf nitrogen content, and specific leaf area. While some studies (Keenan et al., 2013) attributed the increase in carbon storage capacity to the increased water-use efficiency caused by CO_2 fertilization, others found the changing meteorological conditions to be as important (Dunn et al., 2007; Dragoni et al., 2011; Pretzsch et al., 2014). Increasing length of growing season has been linked to enhanced NEP (Baldocchi, 2003; Dragoni et al., 2011), but other studies have found that contradictory effects from increasing annual air and soil temperature have no impact on NEP–productivity can be enhanced during a longer growing season and by temperature rise, but so can respiration (Dunn et al., 2007; Piao et al., 2008; Bond-Lamberty et al., 2018). Changes in precipitation likewise affect the ecosystem carbon balance. When plants experience water stress, they limit photosynthesis to protect their resources. During droughts plants absorb less CO_2, which during dry years is evident in the atmospheric CO_2 growth rate on a global scale (Humphrey et al., 2018).

4.2.2 Nutrients

Phosphorus (P) and nitrogen (N) are the primary nutrients affecting plant growth. Net primary production and decomposition of organic matter in litter and soil are affected by N availability in temperate and boreal ecosystems (Vitousek and Howarth, 1991). From FACE experiments it has been shown that N availability constrains tree response to CO_2 fertilization (Norby et al., 2010). Improving N content stimulates photosynthesis and allocates more C to stem wood and coarser roots, while less carbon is allocated to fine roots resulting in fewer exudates to root symbionts (Janssens and Luyssaert, 2009). Consequently, less microbial mass is contained within the ecosystem which lowers heteorotrophic respiration (Janssens et al., 2010). Together with declines in autotrophic root respiration, this can lead to greater amounts of carbon sequestrated by the soil (Noormets et al., 2015).

From theory and N fertilization experiments, it is evident that the N has a notable impact on the ecosystem's C storage capacity. Omitting the close link between C and N in models can result in erroneous estimates of C stored by forest (Hungate et al., 2003). The lack of a fully coupled N-cycle in global process-based ecosystem models was found to overestimate the carbon sequestered in terrestrial ecosystems by as much as 40-80% for projections made for 1850-2100 (Zaehle et al., 2014). Much effort has therefore been put into including the N-cycle in next generation global ecosystem models (e.g., Goll et al., 2017; Vuichard et al., 2018). Increasing N deposition has been a contributing factor for the observed terrestrial C sink in recent decades (Magnani et al., 2007), while the reduction of N deposition in recent years partially explained the slowdown of stem volume increment in European forests (Nabuurs et al., 2013).

4.2.3 Soil

Disturbances remove living aboveground forest biomass, making soil C sequestration important for the long-term storage capacity. However, there is a scarcity of long-term experiments (spanning more than a decade) capable of capturing the full impact on soil organic carbon from changes in environmental drivers (Hyvnen et al., 2007), and the sensitivity of heterotrophic respiration to changes in temperature, precipitation and inputs of organic matter is highly uncertain. A recent synthesis by Bond-Lamberty et al. (2018) using a global soil respiration database found that soil heterotrophic respiration has risen in recent decades. They proposed two non-mutually exclusive mechanisms to be responsible: increases in GPP resulting in more organic matter being available for microbial metabolism, and enhanced climate-driven mineralization primarily led by increases in temperatures. Similarly, Noormets et al. (2015) found that responses of respiration to temperature are mediated by carbohydrate availability, and are controlled by productivity and allocation, whereas the impact of CO_2 fertilization on soil carbon stocks have been found to be limited.

4.3 Natural disturbances

High severity disturbances can simplify the canopy structure, while low and moderate disturbances enhance canopy complexity (Gough et al., 2016). The high severity of fire disturbances can be catastrophic for an ecosystem and are evident as fire scars in forest age-related patterns across North America (Pan et al., 2011b), although other ecosystems depend on fire for survival. The carbon sequestered by the burned biomass is released into the atmosphere, and wildfires between 1990 and 2016 were estimated to annually emit between 11 and 242 Mt CO_2 eqv per year (Kurz et al., 2018). Extreme weather events, pathogens, insects, and age-related senescence (Gough et al., 2016) fall under the low to moderate severity disturbance events category with some of these possibly intertwined. Bark beetle attacks have, for example, been linked to wind-throw events (Schroeder, 2010; Kärvemo, 2015). These disturbances can open up the canopy changing the distribution of resources of light and nitrogen. This allows for regeneration and diversification of plant functionality, sustaining forest carbon uptake in middle to late successional forests (Gough et al., 2016).

4.4 Anthropogenic disturbances

4.4.1 Land-use and land-cover change

Land-use and land-cover change has been driven by demands from a growing population for food, fibers, and fuels. Deforestation creates major disruption in an ecosystem, and historically deforestation has depleted soil carbon content by 24%, primarily due to oxidation of organic matter and soil erosion (Murty et al., 2002; Smith et al., 2016). In the case of afforestation, the legacy effect of previous management choices controls the soil quality, which in turn can affect both the new forest growth and resilience to climatic changes (Mausolf et al., 2018). Developed countries in boreal and temperate zones experience stable or increased forest cover with afforestation (Food and Agriculture Organization of the United Nations, 2015), which can be seen, for example, in forest age-class patterns across the United States, as westbound settlers cleared the forest to exploit the land for agricultural purposes, which again was abandoned in the early twentieth century and forests were allowed to regrow (Pan et al., 2011b).

4.4.2 Forest management

Traditionally, forest management strategies have aimed at maximizing wood production. Intensive silvicultural systems such as clear-cuts and short rotation length reduce the aboveground carbon storage in comparison to less intensive choices such as single tree selection, shelterwood cutting, and continuous cover (e.g. Puhlick et al., 2016). Regardless of the chosen silvicultural system, aboveground biomass is removed to be used for wood products and energy, and both will eventually be released back into the atmosphere as CO_2. The exact amount can be estimated by conducting life-cycle analyses. Following a major harvest, it can take decades to centuries before the same carbon storage is reached in a forest stand, although carbon benefits achieved through energy and product substitution can be substantial. On a larger scale, evidence suggests that the European managed forests are reaching carbon sink saturation under their current management intensity, age-class distribution, and tree species (Nabuurs et al., 2013). Future projections conducted by Luyssaert et al. (2018) found that between 2010 and 2100 the maximal cumulative increase in the forest carbon storage capacity of European forest is 8.1 P g C, only marginally impacting the atmospheric CO_2 concentration.

5 Are forests sources or sinks of carbon?

5.1 Exploration of the recent global forest sink

Over the past two decades, studies have attempted to estimate carbon stocks and fluxes of forests at the global scale (Pacala et al., 2001; Schimel et al., 2001; Bonan, 2008; Pan et al., 2011a). Evidence shows that forests have recently acted as a carbon sink, especially in temperate and boreal zones (Nabuurs et al., 2003; Ciais et al., 2008). Based on national forest inventory statistics collected across Europe, Ciais et al. (2008) showed that European forests have been accumulating carbon since the 1950s. Six causes were given by the researchers involved to explain their results. Ten years later, we think this is a good opportunity to examine these causes in the light of new literature. The following paragraphs show several important studies which estimate or explore drivers of forest sinks, classified according to each cause presented by Ciais et al. (2008). It is important to note that the causes are not mutually exclusive, and that two causes are often closely related.

As described in Section 4, forest management can impact the forest sink. For Ciais et al. (2008), productivity increases outpacing harvest rates was one of the main causes that explained the C sink of current forests. This assumption was confirmed by the work of Luyssaert et al. (2010), who found that the relatively large net forest biomass productivity in the EU-25 was the result of NPP growing faster than carbon losses over the same period. Forest management, through its impact on harvest, thus appears to be a driver behind the recent European forest carbon sink. Closely related to forest management, Ciais et al. (2008) also proposed that forest aging explained part of the forest sink, as covered in Section 3.1 and Fig. 3. Several studies since have shown that past human and natural disturbances have resulted in young northern temperate forests globally (Pan et al., 2011a; Hicke et al., 2012), confirming the notion that age contributes to the sink.

The gain in NPP evoked by Ciais et al. (2008) is closely related to other well-known effects, namely CO_2 fertilization and nitrogen deposition (Section 3.1). Bellassen et al. (2011) found that 87% of the sink trend is explained by these effects at the regional scale during the period 1950-2000. The fertilization effect was recently reaffirmed by the work of Fernández-Martínez et al. (2017) who found that CO_2 fertilization dominated sulfur deposition in the carbon balance of forests in industrialized regions between 1995 and 2011. Other works identify nitrogen deposition as the primary factor in forest carbon sinks. For example, all four process-based models in Van Oijen et al. (2008) identified the increase in nitrogen supply as the major cause of observed changes in European forest growth during the twentieth century.

Afforestation (c.f. Fig. 4) was also proposed by Ciais et al. (2008) to explain the European forest sink. They argued that in Europe new conifer plantations in the 1970s and 1980s increased forest area leading to additional storage of carbon. This has since been supported by a land-surface model inter-comparison that identified afforestation, along with CO_2 and nitrogen fertilization, as one of the main factors explaining the forest sink (Churkina et al., 2010).

The last process highlighted by Ciais et al. (2008) was the abandonment of human practices that allow litter exportation outside of the forest ecosystem, such as grazing and litter raking. The abandonment of these practices increases fertility of forest soils owing to an increase in nutrient input. Based on a modeling experiment, Gimmi et al. (2013) showed that litter raking in Alpine forests

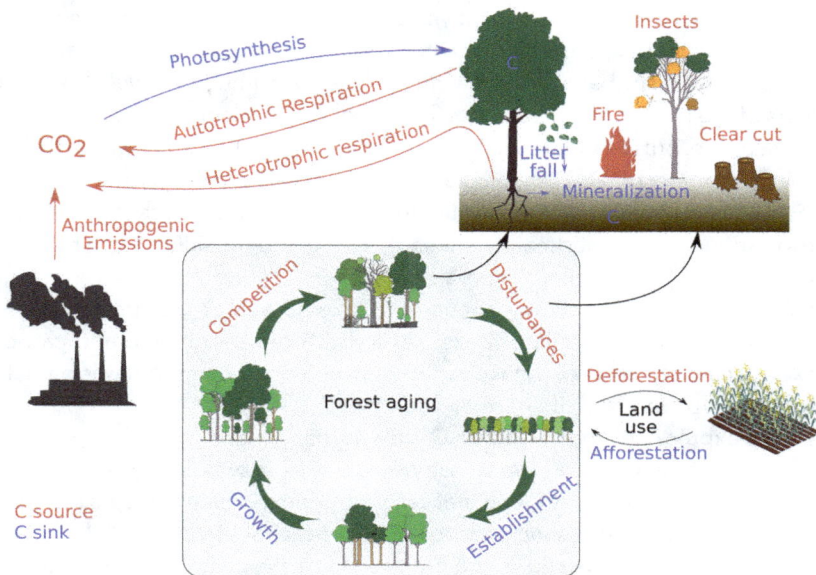

Figure 4 An illustration of influences on forest carbon sinks. Source: modified from Bonan (2008). Reprinted with permission from AAAS.

reduced soil C by an average of 17% after three centuries. These practices were often used in Europe during the last century but were abandoned with the introduction of modern chemical fertilizers. Soils currently act as a C sink as ecosystem processes realign with the contemporary absence of litter raking.

5.2 Future forest carbon sink

The fate of the forest carbon sink under climate change is still under debate. Researchers increasingly tackle this question by running simulations including changing environmental conditions to discern the most likely path. The large majority of these projections showed that temperate and boreal forests will remain a carbon sink for the next century, whether in Europe (De Vries and Posch, 2011; Nabuurs et al., 2013; Lobianco et al., 2016; Valade et al., 2017; Pilli et al., 2017; Pugh et al., 2018), North America (Chen et al., 2000; Hurtt et al., 2002; Metsaranta et al., 2010; Wear and Coulston, 2015), China (Peng et al., 2009; Zhang et al., 2013), or Russia (Dolman et al., 2012). However, simulation-based scenarios depend strongly on both assumptions and processes incorporated into the model. Consequently, these studies often diverge on the magnitude of the sink.

Nabuurs et al. (2013) point out signs of saturation in the capacity of European forests to trap carbon, for three reasons: (1) the total stem volume increment has decreased in recent years; (2) increased land-use pressure has resulted in slower expansion of forested area; and (3) forests are increasingly vulnerable to disturbances. These three reasons summarize well the main drivers of the forest C sink. Note that the authors still anticipate a future forest C sink in Europe, but the rate at which the forests accumulate carbon will decrease. De Vries and Posch (2011) showed an increase in carbon sequestration for the period 2010–50 mainly driven by climate change, unlike the recent past where carbon sequestration was mainly driven by nitrogen deposition. More recently, Pugh et al. (2018) found widespread agreement among multiple vegetation models that boreal forests will continue to take up a large amount of carbon in the long-term, as a result of increases in both climate change and carbon dioxide. The magnitude of uptake varied between simulations and was partially driven by an advance of the northern treeline.

The expression 'forest sector C sink' is often used to encompass the forest ecosystem as well as the forestry sector. This expression includes wood products and energy substitution that can partly compensate a decrease in NEP and slightly change the conclusion of the studies. Valade et al. (2017) found that in 99% of their projections, the European forestry sector remains a strong sink, even if the forests themselves become a source of carbon. Lobianco et al. (2016) found that increases in coniferous mortality and changes in forest growth, in addition to a rise in demand of harvested wood products worldwide, will decrease the average sequestration rate of French forests by 5.8–6.6%. Pilli et al. (2017) used three different scenarios, assuming constant, increasing (+20%) and decreasing (20%) harvest, and afforestation rates compared to the historical period to show that the average annual net sector exchange (i.e. the amount of carbon stored in both the forest and harvested wood products) in Europe will reach 126, 101, and 151 Tg of carbon per year in the year 2030, respectively.

One of the most important factors that can explain the disagreement in model projections for the C sink is the change in natural disturbance regime. Scenarios that account for this result in simulations with a strong slowdown of the C sink (Lindroth et al., 2009; Pan et al., 2011a; Metsaranta et al., 2010; Hicke et al., 2012; Seidl et al., 2014). Estimations of large reductions in the carbon sink after the European storms Lothar and Gudrun show that large wind-throw events may partly explain the large inter-annual variability in the terrestrial carbon sink (Lindroth et al., 2009). In central Europe, Seidl et al. (2014) showed that disturbance intensification scenarios can offset the effect of management strategies aiming to increase the forest carbon sink with an estimated disturbance-related reduction of the forest carbon storage potential of 503.4 Tg C in 2021–30. In the forests of North America, the future of the forest sink seems closely related to wildfire and bark beetle outbreaks. In the United States, recovery from these disturbances represents the primary reason behind the contemporary forest carbon sink, with a significant reduction predicted for the future (Hurtt et al., 2002). On the other hand, Canada's managed forests have already seen the transition from sink to source due to disturbances, with a likely transition back to an overall sink in the next few decades (Metsaranta et al., 2010). A more recent study in the United States finds a significant sink in current US forests with a projected decline in sink strength over the next 25 years (Wear and Coulston, 2015), although the forests will remain a strong sink. The reasons for this decline vary according to region, with disturbances and harvest rates playing key roles.

6 Carbon management as distinct from climate management

The Paris Agreement sets a goal to keep global temperature rise below two degrees compared to pre-industrial levels (Article 2), setting out two primary mechanisms to accomplish it: (1) reducing emissions of greenhouse gases (GHGs; Articles 4 and 6); and (2) increasing sinks and reservoirs of GHGs (Article 4, with Article 5 explicitly for forests) (Member countries of the United Nations, 2015). Recent modeling using a single process-based ecosystem model coupled to a general circulation model demonstrated that maximizing the European forest carbon sink through changes in forest management (i.e., species composition and harvest intensity) did not result in reduction of surface temperature while keeping forest area constant (Luyssaert et al., 2018). An earlier study, focused on reforestation in New England (the United States), showed that adding forest area through conversion from grasslands/croplands to forest generally led to warming (Burakowski et al., 2016). While neither of these studies include possible cooling effects due to cloud formation caused by biogenic volatile organic compound (BVOC) emissions from forests, they still demonstrate the conflict between scientific findings and political action over the past several decades concerning the relationships between carbon, forests, and the climate, showing that land-cover change (changing forest area through afforestation/deforestation) and forest management (changing properties of the forest while leaving the forest area the same) can increase carbon storage without cooling the climate.

Both the Kyoto Protocol and the Paris Agreement focus on emission reduction, including by afforestation and reforestation, as a way to stabilize global climate through temperature reduction (Member countries of the United Nations, 1998, 2015); despite that the word 'temperature' does not appear in the Kyoto Protocol, this is implied by the Protocol existing under the United Nations Framework Convention on Climate Change. The underlying idea is that greenhouse gases emitted to the atmosphere (in particular carbon dioxide) are causing temperatures to increase, and by reversing this process humanity can reverse the effects. Such a result is true under the assumption that everything other than the concentration of greenhouse gases remains the same. Scientists have long been aware that this assumption is not valid when it comes to forests, in particular in the temperate and boreal zones affected by snowfall (see Fig. 5).

Houghton et al. (1990), carrying out a literature review as part of the first IPCC report, focus primarily on carbon cycle impacts of tropical forests, recognizing that deforestation can directly affect climate through biophysical changes but saying the impacts on global mean climate are small. Indeed, much of the literature pre-1990 focused on large-scale deforestation of tropical forests. Manabe and Wetherald (1975) showed that doubling the amount of CO_2 in the atmosphere increases the temperature with a general circulation model. Using this study as an incentive, Woodwell et al. (1983) calculated how much carbon had been released to that point, primarily due to deforestation in the tropics. Shukla et al. (1990) found an increase in temperature and a decrease in precipitation over the Amazon after complete deforestation. Boreal and temperate forests received little attention.

The next three IPCC Assessment Reports treated the issue in a similar manner, focusing on the carbon impacts of afforestation and deforestation in the summary reports for policy makers (IPCC, 1995, 2001, 2007), although

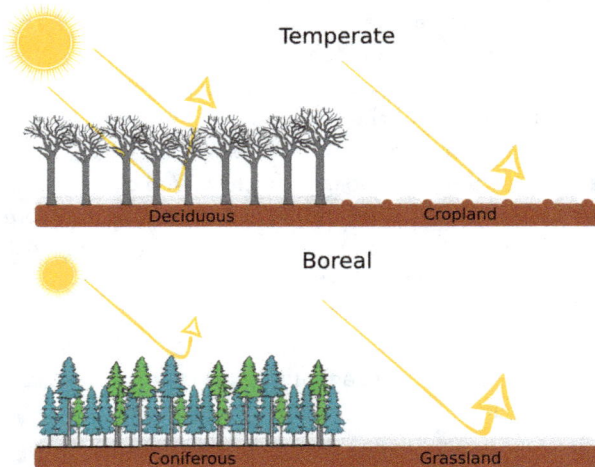

Figure 5 An example of when carbon management does not equal climate management, as storing the same amount of carbon in a coniferous forest and deciduous forests can have opposite climatic impacts due to different modifications of the energy budget. During winter in northern climates, coniferous forests can absorb enough heat to offset the cooling effect of removing carbon dioxide from the atmosphere. Other significant effects, such as evapotranspiration, surface roughness, GHG-balance (including methane and nitrous oxide) and biogenic volatile organic compound (BVOC) emissions are not shown here.

the reports from Working Group I did occasionally touch on possible negative net radiative forcing from deforestation at higher latitudes. The number of studies exploring biophysical impacts of extra-tropical forests continued to grow (Bonan et al., 1992; Foley et al., 1994; Bonan et al., 1995; Lewis, 1998; Betts, 2000; Claussen et al., 2001; Gibbard et al., 2005), generally finding that deforestation at mid to high latitudes would lead to cooling through biophysical impacts. Pielke Sr., et al. (2002) reviewed the physical effects of land-cover change (including forest-cover changes) on climate in addition to the carbon cycle, noting that changing land surface for carbon sequestration under the Kyoto Protocol could have regional and global climatic impacts.

The Fifth Assessment Report from the IPCC (IPCC, 2014) does not deviate from previous reports, calling afforestation, sustainable forest management, and reduced deforestation the most cost-effective mitigation options in the forestry sector, and focusing on the removal of carbon dioxide from the atmosphere. The report does note altered surface reflectance as a side-effect of afforestation. Bala et al. (2007) found that global deforestation results in global cooling, due to biophysical effects overwhelming carbon-cycle warming effects. Brovkin et al. (2009) used an Earth system model to show that global afforestation increased the mean annual surface temperature, with much of the warming being located in the boreal zone. Davin and de Noblet-Ducoudré (2010) showed that replacing global forests by grasslands would cool the climate due to changes in albedo. Arora and Montenegro (2011) used an Earth system model to assess the climate impacts of five different afforestation scenarios, showing that boreal and temperate afforestation resulted in lower temperature reductions than tropical afforestation. Lee et al. (2011) observed higher temperatures in forests than in surrounding open land at high latitudes (north of 45 degrees) using EC methods and local weather stations. Chen et al. (2012) demonstrated the varying timescales of seasonal cooling and warming through afforestation in the southeastern United States. Biophysical effects are generally more important at high latitudes. These results should not be interpreted to mean that deforestation of boreal forests is a preferred method to combat climate change, as such actions would have drastic impacts on other ecosystem services like water supply and quality, in addition to increasing the amount of carbon dioxide in the atmosphere and lowering the pH of the ocean and other water bodies (Halevy and Bachan, 2017).

Since the publication of the fifth IPCC report, additional studies have demonstrated the importance of considering biophysical effects in addition to carbon cycle effects when dealing with forests and climate, in particular outside tropical regions. Naudts et al. (2016) demonstrated that two and a half centuries of forest management in Europe did not cool the continent despite significant carbon sequestration. The authors attribute this primarily to the biophysical properties of tree species, as historical European management converted deciduous trees to coniferous trees. Perugini et al. (2017) provide a recent review on the biophysical effects of land-use change on climate, showing that both boreal and temperate deforestation lead to global cooling, although the number of studies is small; the regional cooling effects are more well-documented.

The IPCC, due to the requirement of treating a very complicated subject and expressing it in ways that will aid decision makers, represent forests as a way to mitigate global warming by reducing atmospheric carbon dioxide concentrations. Such a view is straightforward to apply for high-level decision makers as they work to balance competing priorities of climate management and forest provision of other ecosystem services, such as wood production and recreation. However, current scientific findings presents a more nuanced picture, in particular in the temperate and boreal zones where biophysical effects like changes in surface albedo can counteract climate benefits from carbon sequestration. To make things more challenging, forest-climate studies do not yet routinely include possible cooling effects due to forest emission of BVOCs. Achieving the goal of reducing global surface temperatures therefore relies on incorporation of location-specific forest management policies and more intensive research.

7 Future trends and conclusion

Given the breadth of research techniques being applied to forest carbon research, we limit ourselves to a few examples here, examining them in more detail. We attempt to focus on the boreal and temperate zones when possible, though many of these topics are also relevant for the tropics. In general, future research needs will evolve along with society's priorities. Seeking to improve understanding of the impact of the carbon cycle on the climate may have different needs than seeking to improve the accuracy of carbon reporting budgets. Additional topics of interest include data assimilation into process-based ecosystem models (cf., Schulze et al. (2009); MacBean et al. (2016)), and model representation of natural disturbances like wind (Chen et al., 2018), fire (Rabin et al., 2017), drought (Adams et al., 2017), and insects (Kurz et al., 2008)).

7.1 Operational frameworks

Countries have a pressing need to perform annual reporting and accounting to meet international and national policy objectives, and must compromise to achieve a balance between accuracy and efficiency. For example, models which accurately predict site-scale biomass may not be capable of running across a whole continent, due to lack of detailed data for parameterization or computational resource limitation. At the same time, Luyssaert et al. (2012) searched for largely independent accounting schemes to constrain estimates of GHG balances, using atmospheric inversions, land-based flux measurements, and land-based carbon inventories, combining so-called 'bottom-up' and 'top-down' approaches. Such frameworks, operationalized to give annual reports in language and terms accessible to policy makers, could be run as ecological forecasts (Clark et al., 2001; Dietze et al., 2018) promoting improvement much in the way that weather forecasts have improved model skill and our understanding of the weather system over the past half-century (Bauer et al., 2015).

7.2 LiDAR sampling of tree biomass

As outlined in Section 3.2, accurate and precise estimation of aboveground tree biomass requires significant effort and often destructive sampling techniques. On the other hand, accurate estimation of tree biomass is vital for estimation of current carbon stocks and fluxes (e.g. annual NPP or wood increment). LiDAR provides one promising direction to improve existing allometric relations by increasing the amount of samples which can be collected and tailoring them to regional landscapes. LiDAR allows researchers to create a three-dimensional model of a tree without cutting it down, by reflecting light off various tree parts and calculating distances based on the return time of the beam. Early work involved airborne LiDAR as a form of allometric equation, fitting regressions to measured quantities like height and crown diameter (cf., Nilsson, 1996; Popescu et al., 2003). Stovall et al. (2017) improve on existing on-the-ground methods by allowing more realistic tree shapes when fitting the form of the tree; fitting tree forms is required as the LiDAR data consists of a cloud of spatial points. This model enables a volume to be calculated, which can then be converted into biomass through the use of a species-specific wood density. The results were promising in a Colorado (USA) lodgepole pine forest, though the authors expect more challenges in broadleaf forests.

7.3 Improved sampling networks

FLUXNET represents an estimated 900 active and historical sites across the globe (approximately one-third of which are located in forest ecosystems) reporting flux data (Baldocchi et al., 2001; Chu et al., 2017), including carbon fluxes from EC methods. Networks like FLUXNET provide detailed observational data that are critical to validate models and understand how carbon moves through forests. The networks suffer from several biases, however. Most notable is the geographic distribution, focusing primarily on developed countries in North America and Europe (missing, for example, large swaths of northern boreal forest (Pan et al., 2011a)). In addition, the majority of sites have only been operating for less than five years, limiting temporal representation (Chu et al., 2017). These networks were also developed as more of a harmonization of existing sites as opposed to a tool designed to meet a specific purpose. Instead of examining the network and placing new stations to fill identified gaps, sites were selected for other reasons (Sulkava et al., 2011). Designing more representative networks, and establishing networks primarily to respond to specific scientific questions, would improve the collection of targeted datasets for forest carbon cycles.

7.4 Inclusion of human activities in large-scale ecosystem models

Humans manage a large part of the world's temperate and boreal forests (Erb et al., 2017). In Europe, forests have been managed for hundreds of years (McGrath et al., 2015), which has significant impacts on both current and future carbon stocks. Indeed, researchers partially attribute the current ability of European forests to act as a carbon sink to improvements in forest management practices following the Second World War (Nabuurs et al., 2013). Despite this importance, human management activities in forests are typically not included in Earth system models used to predict future climate, with the exception of land-use change (e.g. converting a forest to grassland or vice-versa) (Pongratz et al., 2018). Luyssaert et al. (2014) used observational evidence to show that human management of forests can result in biophysical changes on the same order of magnitude as complete land-cover conversion. A significant number of Earth system model groups now include at least a minimal harvest model (i.e. removing a prescribed amount of carbon every year) (Pongratz et al., 2018), but much work remains to include a realistic group of processes describing the entirety of harvest impacts, such as disruptions in soil carbon. To compound the issue, such models are frequently parameterized and validated against datasets which include management (e.g. forest inventories, global remote sensing products). Such practices can mask true sources of uncertainty in the models.

8 Acknowledgements

This project has received funding from the European Union's Horizon 2020 research and innovation programme under grant agreement No. 776810 (MJM, PP), No 641816 (CRESCENDO) (ASL), and H2020-MSCA-IF-2016 grant number 747646 (AV). YC was funded through the Ministry of Science and Technology, R.O.C. (MOST 106-2111-M-001-001-MY3).

9 Where to look for further information

9.1 Further reading

- Noormets et al. (2015) give a recent review of forest management and carbon sequestration in global forests, not specific to temperate and boreal zones.

- Chapin III et al. (2006) harmonize terminology around carbon cycle fluxes in forests.
- Curtis and Gough (2018) give a recent overview on forest age and the carbon cycle.
- Thom and Seidl (2016) reviewed the effect of disturbances on temperate and boreal forests.
- Perugini et al. (2017) provide a recent review on the biophysical effects of land-use change on climate.

9.2 Key journals/conferences

In addition to the highest quality general scientific journals (e.g. Nature, Science, PNAS, Global Change Biology), we recommend the following more specialized journals:

- Forest Ecology and Management
- Annals of Forest Science
- Carbon Balance and Management
- Agricultural and Forest Meteorology
- Tree Physiology.

9.3 Major international research projects

- VERIFY, a European project to create an observation-based system for monitoring and verification of greenhouse gases.
- Global Carbon Project, a Global Research Project of Future Earth started in 2001 with the goal to develop a complete picture of the global carbon cycle.
- ICOS, a network of '100 greenhouse gases measuring stations aimed at quantifying and understanding the greenhouse gas balance of the Europe and neighboring regions.'

10 References

Adams, H. D., Zeppel, M. J. B., Anderegg, W. R. L., Hartmann, H., Landhäusser, S. M., Tissue, D. T., Huxman, T. E., Hudson, P. J., Franz, T. E., Allen, C. D., Anderegg, L. D. L., Barron-Gafford, G. A., Beerling, D. J., Breshears, D. D., Brodribb, T. J., Bugmann, H., Cobb, R. C., Collins, A. D., Dickman, L. T., Duan, H., Ewers, B. E., Galiano, L., Galvez, D. A., Garcia-Forner, N., Gaylord, M. L., Germino, M. J., Gessler, A., Hacke, U. G., Hakamada, R., Hector, A., Jenkins, M. W., Kane, J. M., Kolb, T. E., Law, D. J., Lewis, J. D., Limousin, J., Love, D. M., Macalady, A. K., Martínez-Vilalta, J., Mencuccini, M., Mitchell, P. J., Muss, J. D., O'Brien, M. J., O'Grady, A. P., Pangle, R. E., Pinkard, E. A., Piper, F. I., Plaut, J. A., Pockman, W. T., Quirk, J., Reinhardt, K., Ripullone, F., Ryan,

M. G., Sala, A., Sevanto, S., Sperry, J. S., Vargas, R., Vennetier, M., Way, D. A., Xu, C., Yepez, E. A. and McDowell, N. G. 2017. A multi-species synthesis of physiological mechanisms in drought-induced tree mortality. *Nature Ecology and Evolution* 1(9), 1285–91. doi:10.1038/s41559-017-0248-x.

Addo-Danso, S. D., Prescott, C. E. and Smith, A. R. 2016. Methods for estimating root biomass and production in forest and woodland ecosystem carbon studies: a review. *Forest Ecology and Management* 359, 332–51. doi:10.1016/j.foreco.2015.08.015.

Alkama, R. and Cescatti, A. 2016. Biophysical climate impacts of recent changes in global forest cover. *Science* 351(6273), 600–4. doi:10.1126/science.aac8083.

Arino, O., Bicheron, P., Achard, F., Latham, J., Witt, R. and Weber, J.-L. 2008. Globcover: the most detailed portrait of earth. Available at: http://www.esa.int/esapub/bulletin/bulletin136/bul136d_arino.pdf (accessed on June 5, 2019).

Arneth, A., Harrison, S. P., Zaehle, S., Tsigaridis, K., Menon, S., Bartlein, P. J., Feichter, J., Korhola, A., Kulmala, M., O'Donnell, D., Schurgers, G., Sorvari, S. and Vesala, T. 2010. Terrestrial biogeochemical feedbacks in the climate system. *Nature Geoscience* 3(8), 525–32. doi:10.1038/ngeo905.

Arora, V. K. and Montenegro, A. 2011. Small temperature benefits provided by realistic afforestation efforts. *Nature Geoscience* 4(8), 514–8. doi:10.1038/ngeo1182.

Bala, G., Caldeira, K., Wickett, M., Phillips, T. J., Lobell, D. B., Delire, C. and Mirin, A. 2007. Combined climate and carbon-cycle effects of large-scale deforestation. *Proceedings of the National Academy of Sciences of the United States of America* 104(16), 6550–5. doi:10.1073/pnas.0608998104.

Baldocchi, D. D. 2003. Assessing the eddy covariance technique for evaluating carbon dioxide exchange rates of ecosystems: past, present and future. *Global Change Biology* 9(4), 479–92. doi:10.1046/j.1365-2486.2003.00629.x.

Baldocchi, D., Falge, E., Gu, L., Olson, R., Hollinger, D., Running, S., Anthoni, P., Bernhofer, C., Davis, K., Evans, R., Fuentes, J., Goldstein, A., Katul, G., Law, B., Lee, X., Malhi, Y., Meyers, T., Munger, W., Oechel, W., Paw, K. T., Pilegaard, K., Schmid, H. P., Valentini, R., Verma, S., Vesala, T., Wilson, K. and Wofsy, S. 2001. FLUXNET: a new tool to study the temporal and spatial variability of ecosystem-scale carbon dioxide, water vapor, and energy flux densities. *Bulletin of the American Meteorological Society* 82(11), 2415–34. doi:10.1175/1520-0477(2001)082<2415:FANTTS>2.3.CO;2.

Barredo, J. I., Miguel, J. S., Caudullo, G. and Busetto, L. 2012. A European map of living forest biomass and carbon stock: executive report. JRC Scientific and Policy Reports, European Commission.

Bathiany, S., Claussen, M., Brovkin, V., Raddatz, T. and Gayler, V. 2010. Combined biogeo-physical and biogeochemical effects of large-scale forest cover changes in the MPI earth system model. *Biogeosciences* 7(5), 1383–99. doi:10.5194/bg-7-1383-2010.

Bauer, P., Thorpe, A. and Brunet, G. 2015. The quiet revolution of numerical weather prediction. *Nature* 525(7567), 47–55. doi:10.1038/nature14956.

Beets, P. N., Kimberley, M. O., Oliver, G. R., Pearce, S. H., Graham, J. D. and Brandon, A. 2012. Allometric equations for estimating carbon stocks in natural forest in New Zealand. *Forests* 3(3), 818–39. doi:10.3390/f3030818.

Bellassen, V. and Luyssaert, S. 2014. Carbon sequestration: managing forests in uncertain times. *Nature* 506(7487), 153–5. doi:10.1038/506153a.

Bellassen, V., Viovy, N., Luyssaert, S., Le Maire, G., Schelhaas, M.-J. and Ciais, P. 2011. Reconstruction and attribution of the carbon sink of European forests between 1950 and 2000. *Global Change Biology* 17(11), 3274-92. doi:10.1111/j.1365-2486.2011.02476.x.

Bennett, B. M. and Barton, G. A. 2018. The enduring link between forest cover and rainfall: a historical perspective on science and policy discussions. *Forest Ecosystems* 5(1), 5. doi:10.1186/s40663-017-0124-9.

Betts, R. A. 2000. Offset of the potential carbon sink from boreal forestation by decreases in surface albedo. *Nature* 408(6809), 187-90. doi:10.1038/35041545.

Bonan, G. B. 2008. Forests and climate change: forcings, feedbacks, and the climate benefits of forests. *Science* 320(5882), 1444-9. doi:10.1126/science.1155121.

Bonan, G. B., Pollard, D. and Thompson, S. L. 1992. Effects of boreal forest vegetation on global climate. *Nature* 359(6397), 716-8. doi:10.1038/359716a0.

Bonan, G. B., Chapin III, F. S. and Thompson, S. L. 1995. Boreal forest and tundra ecosystems as components of the climate system. *Climatic Change* 29(2), 145-67. doi:10.1007/BF01094014.

Bond-Lamberty, B., Wang, C. and Gower, S. T. 2002. Aboveground and belowground biomass and sapwood area allometric equations for six boreal tree species of northern Manitoba. *Canadian Journal of Forest Research* 32(8), 1441-50. doi:10.1139/x02-063.

Bond-Lamberty, B., Bailey, V. L., Chen, M., Gough, C. M. and Vargas, R. 2018. Globally rising soil heterotrophic respiration over recent decades. *Nature* 560(7716), 80-3. doi:10.1038/s41586-018-0358-x.

Breithaupt, J. L., Smoak, J. M., Smith, T. J., Sanders, C. J. and Hoare, A. 2012. Organic carbon burial rates in mangrove sediments: strengthening the global budget. *Global Biogeochemical Cycles* 26(3), GB3011. doi:10.1029/2012GB004375.

Bright, R. M., Davin, E., O'Halloran, T., Pongratz, J., Zhao, K. and Cescatti, A. 2017. Local temperature response to land cover and management change driven by non-radiative processes. *Nature Climate Change* 7(4), 296-302. doi:10.1038/nclimate3250.

Brovkin, V., Raddatz, T., Reick, C. H., Claussen, M. and Gayler, V. 2009. Global biogeophysical interactions between forest and climate. *Geophysical Research Letters* 36(7), L07405. doi:10.1029/2009GL037543.

Brus, D. J., Hengeveld, G. M., Walvoort, D. J. J., Goedhart, P. W., Heidema, A. H., Nabuurs, G. J. and Gunia, K. 2012. Statistical mapping of tree species over Europe. *European Journal of Forest Research* 131(1), 145-57. doi:10.1007/s10342-011-0513-5.

Burakowski, E. A., Ollinger, S. V., Bonan, G. B., Wake, C. P., Dibb, J. E. and Hollinger, D. Y. 2016. Evaluating the climate effects of reforestation in New England using a weather research and forecasting (WRF) model multiphysics ensemble. *Journal of Climate* 29(14), 5141-56. doi:10.1175/JCLI-D-15-0286.1.

Campioli, M., Vicca, S., Luyssaert, S., Bilcke, J., Ceschia, E., Chapin III, F. S., Ciais, P., Fernández-Martínez, M., Malhi, Y., Obersteiner, M., Olefeldt, D., Papale, D., Piao, S. L., Peñuelas, J., Sullivan, P. F., Wang, X., Zenone, T. and Janssens, I. A. 2015. Biomass production efficiency controlled by management in temperate and boreal ecosystems. *Nature Geoscience* 8(11), 843-6. doi:10.1038/ngeo2553.

Campioli, M., Malhi, Y., Vicca, S., Luyssaert, S., Papale, D., Nuelas, J. P., Reichstein, M., Migliavacca, M., Arain, M. A. and Janssens, I. A. 2017. Evaluating the convergence between eddy-covariance and biometric methods for assessing carbon budgets of forests. *Nature Communications* 7, 13717.

Chapin III, F. S., Woodwell, G. M., Randerson, J. T., Rastetter, E. B., Lovett, G. M., Baldocchi, D. D., Clark, D. A., Harmon, M. E., Schimel, D. S., Valentini, R., Wirth, C., Aber, J. D., Cole, J. J., Goulden, M. L., Harden, J. W., Heimann, M., Howarth, R. W., Matson, P. A., McGuire, A. D., Melillo, J. M., Mooney, H. A., Neff, J. C., Houghton, R. A., Pace, M. L., Ryan, M. G., Running, S. W., Sala, O. E., Schlesinger, W. H. and Schulze, E.-D. 2006. Reconciling carbon-cycle concepts, terminology, and methods. *Ecosystems* 9(7), 1041–50. doi:10.1007/s10021-005-0105-7.

Charru, M., Seynave, I., Hervé, J.-C., Bertrand, R. and Bontemps, J.-D. 2017. Recent growth changes in Western European forests are driven by climate warming and structured across tree species climatic habitats. *Annals of Forest Science* 74(2), 33. doi:10.1007/s13595-017-0626-1.

Chen, W., Chen, J. and Cihlar, J. 2000. An integrated terrestrial ecosystem carbon-budget model based on changes in disturbance, climate, and atmospheric chemistry. *Ecological Modelling* 135(1), 55–79. doi:10.1016/S0304-3800(00)00371-9.

Chen, G. S., Notaro, M., Liu, Z. and Liu, Y. 2012. Simulated local and remote biophysical effects of afforestation over the southeast United States in boreal summer. *Journal of Climate* 25(13), 4511–22. doi:10.1175/JCLI-D-11-00317.1.

Chen, Y.-Y., Gardiner, B., Pasztor, F., Blennow, K., Ryder, J., Valade, A., Naudts, K., Otto, J., McGrath, M. J., Planque, C. and Luyssaert, S. 2018. Simulating damage for wind storms in the land surface model ORCHIDEE-CAN (revision 4262). *Geoscientific Model Development* 11(2), 771–91. doi:10.5194/gmd-11-771-2018.

Chevallier, F., Ciais, P., Conway, T. J., Aalto, T., Anderson, B. E., Bousquet, P., Brunke, E. G., Ciattaglia, L., Esaki, Y., Frhlich, M., Gomez, A., Gomez-Pelaez, A. J., Haszpra, L., Krummel, P. B., Langenfelds, R. L., Leuenberger, M., Machida, T., Maignan, F., Matsueda, H., Morgu, J. A., Mukai, H., Nakazawa, T., Peylin, P., Ramonet, M., Rivier, L., Sawa, Y., Schmidt, M., Steele, L. P., Vay, S. A., Vermeulen, A. T., Wofsy, S. and Worthy, D. 2010. CO_2 surface fluxes at grid point scale estimated from a global 21 year reanalysis of atmospheric measurements. *Journal of Geophysical Research: Atmospheres* 115(D21), D21307. doi:10.1029/2010JD013887.

Chu, H., Baldocchi, D. D., John, R., Wolf, S. and Reichstein, M. 2017. Fluxes all of the time? a primer on the temporal representativeness of FLUXNET. *Journal of Geophysical Research: Biogeosciences* 122(2), 289–307. doi:10.1002/2016JG003576.

Churkina, G., Zaehle, S., Hughes, J., Viovy, N., Chen, Y., Jung, M., Heumann, B. W., Ramankutty, N., Heimann, M. and Jones, C. 2010. Interactions between nitrogen deposition, land cover conversion, and climate change determine the contemporary carbon balance of Europe. *Biogeosciences* 7(9), 2749–64. doi:10.5194/bg-7-2749-2010.

Ciais, P., Schelhaas, M. J., Zaehle, S., Piao, S. L., Cescatti, A., Liski, J., Luyssaert, S., Le-Maire, G., Schulze, E. -D., Bouriaud, O., Freibauer, A., Valentini, R. and Nabuurs, G. J. 2008. Carbon accumulation in European forests. *Nature Geoscience* 1(7), 425–9. doi:10.1038/ngeo233.

Ciais, P., Canadell, J. G., Luyssaert, S., Chevallier, F., Shvidenko, A., Poussi, Z., Jonas, M., Peylin, P., King, A. W., Schulze, E.-D., Piao, S., Rödenbeck, C., Peters, W. and Bréon, F. 2010. Can we reconcile atmospheric estimates of the northern terrestrial carbon sink with land-based accounting? *Current Opinion in Environmental Sustainability* 2(4), 225–30. doi:10.1016/j.cosust.2010.06.008.

Ciais, P., Sabine, C., Bala, G., Bopp, L., Brovkin, V., Canadell, J., Chhabra, A., DeFries, R., Galloway, J., Heimann, M., Jones, C., Le Quéré, C., Myneni, R., Piao, S. and Thornton,

P. 2013. Carbon and other biogeochemical cycles. In: *Climate Change 2013: the Physical Science Basis. Contribution of Working Group I to the Fifth Assessment Report of the Intergovernmental Panel on Climate Change.* Cambridge University Press, Cambridge, United Kingdom and New York, NY, pp. 465–570.

Clark, J. S., Carpenter, S. R., Barber, M., Collins, S., Dobson, A., Foley, J. A., Lodge, D. M., Pascual, M., Pielke, R., Pizer, W., Pringle, C., Reid, W. V., Rose, K. A., Sala, O., Schlesinger, W. H., Wall, D. H. and Wear, D. 2001. Ecological forecasts: an emerging imperative. *Science* 293(5530), 657–60. doi:10.1126/science.293.5530.657.

Claussen, M., Brovkin, V. and Ganopolski, A. 2001. Biogeophysical versus biogeochemical feedbacks of large-scale land cover change. *Geophysical Research Letters* 28(6), 1011–4. doi:10.1029/2000GL012471.

Coomes, D. A., Šafka, D., Shepherd, J., Dalponte, M. and Holdaway, R. 2018. Airborne laser scanning of natural forests in New Zealand reveals the influences of wind on forest carbon. *Forest Ecosystems* 5(1), 10.

Crowther, T. W., Glick, H. B., Covey, K. R., Bettigole, C., Maynard, D. S., Thomas, S. M., Smith, J. R., Hintler, G., Duguid, M. C., Amatulli, G., Tuanmu, M. N., Jetz, W., Salas, C., Stam, C., Piotto, D., Tavani, R., Green, S., Bruce, G., Williams, S. J., Wiser, S. K., Huber, M. O., Hengeveld, G. M., Nabuurs, G. J., Tikhonova, E., Borchardt, P., Li, C. F., Powrie, L. W., Fischer, M., Hemp, A., Homeier, J., Cho, P., Vibrans, A. C., Umunay, P. M., Piao, S. L., Rowe, C. W., Ashton, M. S., Crane, P. R. and Bradford, M. A. 2015. Mapping tree density at a global scale. *Nature* 525(7568), 201–5. doi:10.1038/nature14967.

Curtis, P. S. and Gough, C. M. 2018. Forest aging, disturbance and the carbon cycle. *The New Phytologist* 219(4), 1188–93. doi:10.1111/nph.15227.

Davin, E. L. and de Noblet-Ducoudré, N. 2010. Climatic impact of global-scale deforestation: radiative versus nonradiative processes. *Journal of Climate* 23(1), 97–112. doi:10.1175/2009JCLI3102.1.

De Vries, W. and Posch, M. 2011. Modelling the impact of nitrogen deposition, climate change and nutrient limitations on tree carbon sequestration in Europe for the period 1900–2050. *Environmental Pollution* 159(10), 2289–99. doi:10.1016/j.envpol.2010.11.023.

Dietze, M. C., Fox, A., Beck-Johnson, L. M., Betancourt, J. L., Hooten, M. B., Jarnevich, C. S., Keitt, T. H., Kenney, M. A., Laney, C. M., Larsen, L. G., Loescher, H. W., Lunch, C. K., Pijanowski, B. C., Randerson, J. T., Read, E. K., Tredennick, A. T., Vargas, R., Weathers, K. C. and White, E. P. 2018. Iterative near-term ecological forecasting: needs, opportunities, and challenges. *Proceedings of the National Academy of Sciences of the United States of America* 115(7), 1424–32. doi:10.1073/pnas.1710231115.

Dolman, A. J., Shvidenko, A., Schepaschenko, D., Ciais, P., Tchebakova, N., Chen, T., Van Der Molen, M. K., Belelli Marchesini, L., Maximov, T. C., Maksyutov, S. and Schulze, E.-D. 2012. An estimate of the terrestrial carbon budget of Russia using inventory-based, eddy covariance and inversion methods. *Biogeosciences* 9(12), 5323–40. doi:10.5194/bg-9-5323-2012.

Dragoni, D., Schmid, H. P., Wayson, C. A., Potter, H., Grimmond, C. S. B. and Randolph, J. C. 2011. Evidence of increased net ecosystem productivity associated with a longer vegetated season in a deciduous forest in south-central Indiana, USA. *Global Change Biology* 17(2), 886–97. doi:10.1111/j.1365-2486.2010.02281.x.

Dunn, A. L., Barford, C. C., Wofsy, S. C., Goulden, M. L. and Daube, B. C. 2007. A long-term record of carbon exchange in a boreal black spruce forest: means, responses to interannual variability, and decadal trends. *Global Change Biology* 13(3), 577–90. doi:10.1111/j.1365-2486.2006.01221.x.

Erb, K. H., Luyssaert, S., Meyfroidt, P., Pongratz, J., Don, A., Kloster, S., Kuemmerle, T., Fetzel, T., Fuchs, R., Herold, M., Haberl, H., Jones, C. D., Marn-Spiotta, E., McCallum, I., Robertson, E., Seufert, V., Fritz, S., Valade, A., Wiltshire, A. and Dolman, A. J. 2017. Land management: data availability and process understanding for global change studies. *Global Change Biology* 23(2), 512–33. doi:10.1111/gcb.13443.

European Academies Science Advisory Council. 2017. *Multi-Functionality and Sustainability in the European Unions Forests*. DVZ-Daten-Service GmbH, Halle/Saale, Germany.

European Academies Science Advisory Council. 2018. *Negative Emission Technologies: What Role in Meeting Paris Agreement Targets?* Schaefer Druck und Verlag GmbH, Teutschenthal, Germany.

Farquhar, G. D., von Caemmerer, S. and Berry, J. A. 1980. A biochemical model of photosynthetic CO_2 assimilation in leaves of C3 species. *Planta* 149(1), 78–90. doi:10.1007/BF00386231.

Fernández-Martínez, M., Vicca, S., Janssens, I. A., Ciais, P., Obersteiner, M., Bartrons, M., Sardans, J., Verger, A., Canadell, J. G., Chevallier, F., Wang, X., Bernhofer, C., Curtis, P. S., Gianelle, D., Grünwald, T., Heinesch, B., Ibrom, A., Knohl, A., Laurila, T., Law, B. E., Limousin, J. M., Longdoz, B., Loustau, D., Mammarella, I., Matteucci, G., Monson, R. K., Montagnani, L., Moors, E. J., Munger, J. W., Papale, D., Piao, S. L. and Peñuelas, J. 2017. Atmospheric deposition, CO_2, and change in the land carbon sink. *Scientific Reports* 7(1), 9632. doi:10.1038/s41598-017-08755-8.

Foley, J. A., Kutzbach, J. E., Coe, M. T. and Levis, S. 1994. Feedbacks between climate and boreal forests during the Holocene epoch. *Nature* 371(6492), 52–4. doi:10.1038/371052a0.

Food and Agriculture Organization of the United Nations. 2015. Global forest resources assessment 2015: how are the worlds forests changing? Available at: http://www.fao.org/forest-resources-assessment/past-assessments/fra-2015/en/.

Ford, S. E. and Keeton, W. S. 2017. Enhanced carbon storage through management for old-growth characteristics in northern hardwood-conifer forests. *Ecosphere* 8(4), e01721. doi:10.1002/ecs2.1721.

Gasser, T. and Ciais, P. 2013. A theoretical framework for the net land-to-atmosphere CO_2 flux and its implications in the definition of 'emissions from land-use change'. *Earth System Dynamics* 4(1), 171–86. doi:10.5194/esd-4-171-2013.

Gerbig, C., Lin, J. C., Wofsy, S. C., Daube, B. C., Andrews, A. E., Stephens, B. B., Bakwin, P. S. and Grainger, C. A. 2003. Toward constraining regional-scale fluxes of CO_2 with atmospheric observations over a continent: 2. analysis of COBRA data using a receptor-oriented framework. *Journal of Geophysical Research: Atmospheres* 108(D24), 4757. doi:10.1029/2003JD003770.

Gibbard, S., Caldeira, K., Bala, G., Phillips, T. J. and Wickett, M. 2005. Climate effects of global land cover change. *Geophysical Research Letters* 32(23), L23705. doi:10.1029/2005GL024550.

Gimmi, U., Poulter, B., Wolf, A., Portner, H., Weber, P. and Bürgi, M. 2013. Soil carbon pools in Swiss forests show legacy effects from historic forest litter raking. *Landscape Ecology* 28(5), 835–46. doi:10.1007/s10980-012-9778-4.

Goetz, S. J., Baccini, A., Laporte, N. T., Johns, T., Walker, W., Kellndorfer, J., Houghton, R. A. and Sun, M. 2009. Mapping and monitoring carbon stocks with satellite observations: a comparison of methods. *Carbon Balance and Management* 4(1), 2. doi:10.1186/1750-0680-4-2.

Goll, D. S., Vuichard, N., Maignan, F., Jornet-Puig, A., Sardans, J., Violette, A., Peng, S., Sun, Y., Kvakic, M., Guimberteau, M., Guenet, B., Zaehle, S., Penuelas, J., Janssens, I. and Ciais, P. 2017. A representation of the phosphorus cycle for ORCHIDEE (revision 4520). *Geoscientific Model Development* 10(10), 3745–70. doi:10.5194/gmd-10-3745-2017.

Gonsamo, A., Chen, J. M., Colombo, S. J., Ter-Mikaelian, M. T. and Chen, J. 2017. Global change induced biomass growth offsets carbon released via increased forest fire and respiration of the central Canadian Boreal forest. *Journal of Geophysical Research: Biogeosciences* 122(5), 1275–93. doi:10.1002/2016JG003627.

Gordon, H., Kirkby, J., Baltensperger, U., Bianchi, F., Breitenlechner, M., Curtius, J., Dias, A., Dommen, J., Donahue, N. M., Dunne, E. M., Duplissy, J., Ehrhart, S., Flagan, R. C., Frege, C., Fuchs, C., Hansel, A., Hoyle, C. R., Kulmala, M., Kürten, A., Lehtipalo, K., Makhmutov, V., Molteni, U., Rissanen, M. P., Stozkhov, Y., Tröstl, J., Tsagkogeorgas, G., Wagner, R., Williamson, C., Wimmer, D., Winkler, P. M., Yan, C. and Carslaw, K. S. 2017. Causes and importance of new particle formation in the present-day and preindustrial atmospheres. *Journal of Geophysical Research: Atmospheres* 122(16), 8739–60. doi:10.1002/2017JD026844.

Gordon, C. E., Bendall, E. R., Stares, M. G., Collins, L. and Bradstock, R. A. 2018. Aboveground carbon sequestration in dry temperate forests varies with climate not fire regime. *Global Change Biology* 24(9), 4280–92. doi:10.1111/gcb.14308.

Gough, C. M., Curtis, P. S., Hardiman, B. S., Scheuermann, C. M. and Bond-Lamberty, B. 2016. Disturbance, complexity, and succession of net ecosystem production in North America's temperate deciduous forests. *Ecosphere* 7(6), e01375. doi:10.1002/ecs2.1375.

Grassi, G., Pilli, R., House, J., Federici, S. and Kurz, W. A. 2018. Science-based approach for credible accounting of mitigation in managed forests. *Carbon Balance and Management* 13(1), 8. doi:10.1186/s13021-018-0096-2.

Gurney, K. R., Law, R. M., Denning, A. S., Rayner, P. J., Pak, B. C., Baker, D., Bousquet, P., Bruhwiler, L., Chen, Y.-H., Ciais, P., Fung, I. Y., Heimann, M., John, J., Maki, T., Maksyutov, S., Peylin, P., Prather, M. and Taguchi, S. 2004. Transcom 3 inversion intercomparison: model mean results for the estimation of seasonal carbon sources and sinks. *Global Biogeochemical Cycles* 18(1), GB1010. doi:10.1029/2003GB002111.

Gutiérrez, A. G., Snell, R. S. and Bugmann, H. 2016. Using a dynamic forest model to predict tree species distributions. *Global Ecology and Biogeography* 25(3), 347–58. doi:10.1111/geb.12421.

Halevy, I. and Bachan, A. 2017. The geologic history of seawater pH. *Science* 355(6329), 1069–71. doi:10.1126/science.aal4151.

Hamilton, J. G., DeLucia, E. H., George, K., Naidu, S. L., Finzi, A. C. and Schlesinger, W. H. 2002. Forest carbon balance under elevated CO_2. *Oecologia* 131(2), 250–60. doi:10.1007/s00442-002-0884-x.

Hansen, M. C., Potapov, P. V., Moore, R., Hancher, M., Turubanova, S. A., Tyukavina, A., Thau, D., Stehman, S. V., Goetz, S. J., Loveland, T. R., Kommareddy, A., Egorov, A., Chini, L., Justice, C. O. and Townshend, J. R. G. 2013. High-resolution global maps of 21st-century forest cover change. *Science* 342(6160), 850–3. doi:10.1126/science.1244693.

Hansis, E., Davis, S. J. and Pongratz, J. 2015. Relevance of methodological choices for accounting of land use change carbon fluxes. *Global Biogeochemical Cycles* 29(8), 1230–46. doi:10.1002/2014GB004997.

Hardiman, B. S., Gough, C. M., Halperin, A., Hofmeister, K. L., Nave, L. E., Bohrer, G. and Curtis, P. S. 2013. Maintaining high rates of carbon storage in old forests: a mechanism linking canopy structure to forest function. *Forest Ecology and Management* 298, 111–9. doi:10.1016/j.foreco.2013.02.031.

Harper, A. B., Cox, P. M., Friedlingstein, P., Wiltshire, A. J., Jones, C. D., Sitch, S., Mercado, L. M., Groenendijk, M., Robertson, E., Kattge, J., Bönisch, G., Atkin, O. K., Bahn, M., Cornelissen, J., Niinemets, Ü, Onipchenko, V., Peñuelas, J., Poorter, L., Reich, P. B., Soudzilovskaia, N. A. and Bodegom, Pv. 2016. Improved representation of plant functional types and physiology in the joint uk land environment simulator (JULES v4.2) using plant trait information. *Geoscientific Model Development* 9(7), 2415–40. doi:10.5194/gmd-9-2415-2016.

Harris, N. L., Goldman, E., Gabris, C., Nordling, J., Minnemeyer, S., Ansari, S., Lippmann, M., Bennett, L., Raad, M., Hansen, M. and Potapov, P. 2017. Using spatial statistics to identify emerging hot spots of forest loss. *Environmental Research Letters* 12(2).

He, L., Chen, J. M., Pan, Y., Birdsey, R. and Kattge, J. 2012. Relationships between net primary productivity and forest stand age in U.S. forests. *Global Biogeochemical Cycles* 26(3), GB3009. doi:10.1029/2010GB003942.

Hicke, J. A., Allen, C. D., Desai, A. R., Dietze, M. C., Hall, R. J., Hogg, E. H. T., Kashian, D. M., Moore, D., Raffa, K. F., Sturrock, R. N. and Vogelmann, J. 2012. Effects of biotic disturbances on forest carbon cycling in the United States and Canada. *Global Change Biology* 18(1), 7–34. doi:10.1111/j.1365-2486.2011.02543.x.

Houghton, R. A., Hobbie, J. E., Melillo, J. M., Moore, B., Peterson, B. J., Shaver, G. R. and Woodwell, G. M. 1983. Changes in the carbon content of terrestrial biota and soils between 1860 and 1980: a net release of CO_2 to the atmosphere. *Ecological Monographs* 53(3), 235–62. doi:10.2307/1942531.

Houghton, J. T., Jenkins, G. J. and Ephraums, J. J. (Eds). 1990. *Climate Change: the IPCC Scientific Assessment*. Cambridge University Press, New York, NY.

Humphrey, V., Zscheischler, J., Ciais, P., Gudmundsson, L., Sitch, S. and Seneviratne, S. I. 2018. Sensitivity of atmospheric CO_2 growth rate to observed changes in terrestrial water storage. *Nature* 560(7720), 628–31. doi:10.1038/s41586-018-0424-4.

Hungate, B. A., Dukes, J. S., Shaw, M. R., Luo, Y. and Field, C. B. 2003. Atmospheric science. Nitrogen and climate change. *Science* 302(5650), 1512–3. doi:10.1126/science.1091390.

Hurtt, G. C., Pacala, S. W., Moorcroft, P. R., Caspersen, J., Shevliakova, E., Houghton, R. A. and Moore, B. 2002. Projecting the future of the US carbon sink. *Proceedings of the National Academy of Sciences* 99(3), 1389–94. doi:10.1073/pnas.012249999.

Hyvönen, R., Ågren, G. I., Linder, S., Persson, T., Cotrufo, M. F., Ekblad, A., Freeman, M., Grelle, A., Janssens, I. A., Jarvis, P. G., Kellomäki, S., Lindroth, A., Loustau, D., Lundmark, T., Norby, R. J., Oren, R., Pilegaard, K., Ryan, M. G., Sigurdsson, B. D., Strömgren, M., van Oijen, M. and Wallin, G. 2007. The likely impact of elevated [CO_2], nitrogen deposition, increased temperature and management on carbon sequestration in temperate and boreal forest ecosystems: a literature review. *The New Phytologist* 173(3), 463–80. doi:10.1111/j.1469-8137.2007.01967.x.

IPCC. 1995. *IPCC Second Assessment: Climate Change 1995*.

IPCC. 2001. *Climate Change 2001: Synthesis Report*. Cambridge University Press, Cambridge, UK.

IPCC. 2007. *Climate Change 2007: Synthesis Report*. IPCC, Geneva, Switzerland.

IPCC. 2014. *Climate Change 2014: Synthesis Report*. IPCC, Geneva, Switzerland.

IPCC. 2018. *Global Warming of 1.5 ◦ C: Summary for Policymakers*.

Janssens, I. A. and Luyssaert, S. 2009. Carbon cycle: nitrogen's carbon bonus. *Nature Geoscience* 2(5), 318-9. doi:10.1038/ngeo505.

Janssens, I. A., Dieleman, W., Luyssaert, S., Subke, J.-A., Reichstein, M., Ceulemans, R., Ciais, P., Dolman, A. J., Grace, J., Matteucci, G., Papale, D., Piao, S. L., Schulze, E., Tang, J. and Law, B. E. 2010. Reduction of forest soil respiration in response to nitrogen deposition. *Nature Geoscience* 3(5), 315-22. doi:10.1038/ngeo844.

Jenkins, J. C., Chojnacky, D. C., Heath, L. S. and Birdsey, R. A. 2003. National scale biomass estimators for United States tree species. *Forest Science* 49, 12-35.

Jung, M., Maire, G. L., Zaehle, S., Luyssaert, S., Vetter, M., Churkina, G., Ciais, P., Viovy, N. and Reichstein, M. 2007. Assessing the ability of three land ecosystem models to simulate gross carbon uptake of forests from boreal to Mediterranean climate in Europe. *Biogeosciences* 4(4), 647-56. doi:10.5194/bg-4-647-2007.

Karst, J., Gaster, J., Wiley, E. and Landhäusser, S. M. 2017. Stress differentially causes roots of tree seedlings to exude carbon. *Tree Physiology* 37(2), 154-64. doi:10.1093/treephys/tpw090.

Kärvemo, S. 2015. Outbreak dynamics of the spruce bark beetle Ips typographus in time and space. PhD thesis. Swedish University of Agricultural Sciences.

Keenan, T. F., Hollinger, D. Y., Bohrer, G., Dragoni, D., Munger, J. W., Schmid, H. P. and Richardson, A. D. 2013. Increase in forest water-use efficiency as atmospheric carbon dioxide concentrations rise. *Nature* 499(7458), 324-7. doi:10.1038/nature12291.

Kull, S. J., Rampley, G., Morken, S., Metsaranta, J., Neilson, E. T. and Kurz, W. 2016. *Operational-Scale Carbon Budget Model of the Canadian Forest Sector (CBM-CFS3) Version 1.2: Users Guide*. Natural Resources Canada, Edmonton.

Kurz, W. A., Dymond, C. C., Stinson, G., Rampley, G. J., Neilson, E. T., Carroll, A. L., Ebata, T. and Safranyik, L. 2008. Mountain pine beetle and forest carbon feedback to climate change. *Nature* 452(7190), 987-90. doi:10.1038/nature06777.

Kurz, W. A., Hayne, S., Fellows, M., MacDonald, J. D., Metsaranta, J. M., Hafer, M. and Blain, D. 2018. Quantifying the impacts of human activities on reported greenhouse gas emissions and removals in Canada's managed forest: conceptual framework and implementation. *Canadian Journal of Forest Research* 48(10), 1227-40. doi:10.1139/cjfr-2018-0176.

Lasch-Born, P., Suckow, F., Gutsch, M., Reyer, C., Hauf, Y., Murawski, A. and Pilz, T. 2015. Forests under climate change: potential risks and opportunities. *Meteorologische Zeitschrift* 24(2), 157-72. doi:10.1127/metz/2014/0526.

Lee, X., Goulden, M. L., Hollinger, D. Y., Barr, A., Black, T. A., Bohrer, G., Bracho, R., Drake, B., Goldstein, A., Gu, L., Katul, G., Kolb, T., Law, B. E., Margolis, H., Meyers, T., Monson, R., Munger, W., Oren, R., Paw U, K. T., Richardson, A. D., Schmid, H. P., Staebler, R., Wofsy, S. and Zhao, L. 2011. Observed increase in local cooling effect of deforestation at higher latitudes. *Nature* 479(7373), 384-7. doi:10.1038/nature10588.

Le Quéré, C., Andrew, R. M., Friedlingstein, P., Sitch, S., Pongratz, J., Manning, A. C., Korsbakken, J. I., Peters, G. P., Canadell, J. G., Jackson, R. B., Boden, T. A., Tans, P. P., Andrews, O. D., Arora, V. K., Bakker, D. C. E., Barbero, L., Becker, M., Betts, R. A., Bopp, L., Chevallier, F., Chini, L. P., Ciais, P., Cosca, C. E., Cross, J., Currie, K., Gasser, T.,

Harris, I., Hauck, J., Haverd, V., Houghton, R. A., Hunt, C. W., Hurtt, G., Ilyina, T., Jain, A. K., Kato, E., Kautz, M., Keeling, R. F., Klein Goldewijk, K., Körtzinger, A., Landschützer, P., Lefèvre, N., Lenton, A., Lienert, S., Lima, I., Lombardozzi, D., Metzl, N., Millero, F., Monteiro, P. M. S., Munro, D. R., Nabel, J. E. M. S., Nakaoka, S., Nojiri, Y., Padin, X. A., Peregon, A., Pfeil, B., Pierrot, D., Poulter, B., Rehder, G., Reimer, J., Rödenbeck, C., Schwinger, J., Séférian, R., Skjelvan, I., Stocker, B. D., Tian, H., Tilbrook, B., Tubiello, F. N., van der Laan-Luijkx, I. T., van der Werf, G. R., van Heuven, S., Viovy, N., Vuichard, N., Walker, A. P., Watson, A. J., Wiltshire, A. J., Zaehle, S. and Zhu, D. 2018. Global carbon budget 2017. *Earth System Science Data* 10(1), 405–48. doi:10.5194/essd-10-405-2018.

Lewis, T. 1998. The effect of deforestation on ground surface temperatures. *Global and Planetary Change* 18(1–2), 1–13. doi:10.1016/S0921-8181(97)00011-8.

Lindner, M., Maroschek, M., Netherer, S., Kremer, A., Barbati, A., Garcia-Gonzalo, J., Seidl, R., Delzon, S., Corona, P., Kolström, M., Lexer, M. J. and Marchetti, M. 2010. Climate change impacts, adaptive capacity, and vulnerability of European forest ecosystems. *Forest Ecology and Management* 259(4), 698–709. doi:10.1016/j.foreco.2009.09.023.

Lindroth, A., Lagergren, F., Grelle, A., Klemedtsson, L., Langvall, O., Weslien, P. and Tuulik, J. 2009. Storms can cause Europe-wide reduction in forest carbon sink. *Global Change Biology* 15(2), 346–55. doi:10.1111/j.1365-2486.2008.01719.x.

Lobianco, A., Caurla, S., Delacote, P. and Barkaoui, A. 2016. Carbon mitigation potential of the French forest sector under threat of combined physical and market impacts due to climate change. *Journal of Forest Economics* 23, 4–26. doi:10.1016/j.jfe.2015.12.003.

Luyssaert, S., Schulze, E. D., Börner, A., Knohl, A., Hessenmöller, D., Law, B. E., Ciais, P. and Grace, J. 2008. Old-growth forests as global carbon sinks. *Nature* 455(7210), 213–5. doi:10.1038/nature07276.

Luyssaert, S., Ciais, P., Piao, S. L., Schulwe, E. -D., Jung, M., Zaehle, S., Schelhaas, M. J., Reichstein, M., Churkina, G., Papale, D., Abril, G., Beer, C., Grace, J., Loustau, D., Matteucci, G., Magnani, F., Nabuurs, G. J., Verbeeck, H., Sulkava, M., van der Werf, G. R. and Janssens, I. A. 2010. The European carbon balance. Part 3: forests. *Global Change Biology* 16(5), 1429–50. doi:10.1111/j.1365-2486.2009.02056.x.

Luyssaert, S., Abril, G., Andres, R., Bastviken, D., Bellassen, V., Bergamaschi, P., Bousquet, P., Chevallier, F., Ciais, P., Corazza, M., Dechow, R., Erb, K. -H., Etiope, G., Fortems-Cheiney, A., Grassi, G., Hartmann, J., Jung, M., Lathière, J., Lohila, A., Mayorga, E., Moosdorf, N., Njakou, D. S., Otto, J., Papale, D., Peters, W., Peylin, P., Raymond, P., Rödenbeck, C., Saarnio, S., Schulze, E. -D., Szopa, S., Thompson, R., Verkerk, P. J., Vuichard, N., Wang, R., Wattenbach, M. and Zaehle, S. 2012. The European land and inland water CO_2, CO, CH_4 and N_2O balance between 2001 and 2005. *Biogeosciences* 9(8), 3357–80. doi:10.5194/bg-9-3357-2012.

Luyssaert, S., Jammet, M., Stoy, P. C., Estel, S., Pongratz, J., Ceschia, E., Churkina, G., Don, A., Erb, K., Ferlicoq, M., Gielen, B., Grünwald, T., Houghton, R. A., Klumpp, K., Knohl, A., Kolb, T., Kuemmerle, T., Laurila, T., Lohila, A., Loustau, D., McGrath, M. J., Meyfroidt, P., Moors, E. J., Naudts, K., Novick, K., Otto, J., Pilegaard, K., Pio, C. A., Rambal, S., Rebmann, C., Ryder, J., Suyker, A. E., Varlagin, A., Wattenbach, M. and Dolman, A. J. 2014. Land management and land-cover change have impacts of similar magnitude on surface temperature. *Nature Climate Change* 4(5), 389–93. doi:10.1038/nclimate2196.

Luyssaert, S., Marie, G., Valade, A., Chen, Y. Y., Djomo, S. N., Ryder, J., Otto, J., Naudts, K., Lans, A. S., Ghattas, J. and McGrath, M. J. 2018. Trade-offs in using European forests to meet climate objectives. *Nature* 562(7726), 259-62. doi:10.1038/s41586-018-0577-1.

Maaroufi, N. I., Nordin, A., Palmqvist, K. and Gundale, M. J. 2016. Chronic nitrogen deposition has a minor effect on the quantity and quality of aboveground litter in a boreal forest. *PLoS One* 11(8), 1-16. doi:10.1371/journal.pone.0162086.

MacBean, N., Peylin, P., Chevallier, F., Scholze, M. and Schürmann, G. 2016. Consistent assimilation of multiple data streams in a carbon cycle data assimilation system. *Geoscientific Model Development* 9(10), 3569-88. doi:10.5194/gmd-9-3569-2016.

MacBean, N., Maignan, F., Bacour, C., Lewis, P., Peylin, P., Guanter, L., Köhler, P., Gómez-Dans, J. and Disney, M. 2018. Strong constraint on modelled global carbon uptake using solar-induced chlorophyll fluorescence data. *Scientific Reports* 8(1), 1973. doi:10.1038/s41598-018-20024-w.

Magnani, F., Mencuccini, M., Borghetti, M., Berbigier, P., Berninger, F., Delzon, S., Grelle, A., Hari, P., Jarvis, P. G., Kolari, P., Kowalski, A. S., Lankreijer, H., Law, B. E., Lindroth, A., Loustau, D., Manca, G., Moncrieff, J. B., Rayment, M., Tedeschi, V., Valentini, R. and Grace, J. 2007. The human footprint in the carbon cycle of temperate and boreal forests. *Nature* 447(7146), 848-50. doi:10.1038/nature05847.

Manabe, S. and Wetherald, R. T. 1975. The effects of doubling the CO_2 concentration on the climate of a general circulation model. *Journal of the Atmospheric Sciences* 32(1), 3-15. doi:10.1175/1520-0469(1975)032<0003:TEODTC>2.0.CO;2.

Matross, D. M., Andrews, A., Pathmathevan, M., Gerbig, C., Lin, J. C., Wofsy, S. C., Daube, B. C., Gottlieb, E. W., Chow, V. Y., Lee, J. T., Zhao, C., Bakwin, P. S., Munger, J. W. and Hollinger, D. Y. 2006. Estimating regional carbon exchange in new England and Quebec by combining atmospheric, ground-based and satellite data. *Tellus B: Chemical and Physical Meteorology* 58(5), 344-58.

Mausolf, K., Haerdtle, W., Jansen, K., Delory, B. M., Hertel, D., Leuschner, C., Temperton, V. M., von Oheimb, G. and Fichtner, A. 2018. Legacy effects of land-use modulate tree growth responses to climate extremes. *Oecologia* 187(3), 825-37. doi:10.1007/s00442-018-4156-9.

McGrath, M. J., Luyssaert, S., Meyfroidt, P., Kaplan, J. O., Bürgi, M., Chen, Y., Erb, K., Gimmi, U., McInerney, D., Naudts, K., Otto, J., Pasztor, F., Ryder, J., Schelhaas, M.-J. and Valade, A. 2015. Reconstructing European forest management from 1600 to 2010. *Biogeosciences* 12(14), 4291-316. doi:10.5194/bg-12-4291-2015.

McGroddy, M. E., Daufresne, T. and Hedin, L. O. 2004. Scaling of C:N:P stoichiometry in forests worldwide: implications of terrestrial redfield-type ratios. *Ecology* 85(9), 2390-401. doi:10.1890/03-0351.

McMahon, S. M., Parker, G. G. and Miller, D. R. 2010. Evidence for a recent increase in forest growth. *Proceedings of the National Academy of Sciences of the United States of America* 107(8), 3611-5. doi:10.1073/pnas.0912376107.

Melillo, J. M., Butler, S., Johnson, J., Mohan, J., Steudler, P., Lux, H., Burrows, E., Bowles, F., Smith, R., Scott, L., Vario, C., Hill, T., Burton, A., Zhou, Y. M. and Tang, J. 2011. Soil warming, carbon-nitrogen interactions, and forest carbon budgets. *Proceedings of the National Academy of Sciences of the United States of America* 108(23), 9508-12. doi:10.1073/pnas.1018189108.

Melillo, J. M., Frey, S. D., DeAngelis, K. M., Werner, W. J., Bernard, M. J., Bowles, F. P., Pold, G., Knorr, M. A. and Grandy, A. S. 2017. Long-term pattern and magnitude of soil carbon feedback to the climate system in a warming world. *Science* 358(6359), 101-5. doi:10.1126/science.aan2874.

Member countries of the United Nations. 1998. *Kyoto Protocol to the United Nations Framework Convention on Climate Change.*

Member countries of the United Nations. 2015. *Paris Agreement.* Available at: https://un fccc.int/sites/default/files/english_paris_agreement.pdf (accessed on June 5, 2019.).

Metsaranta, J., Kurz, W., Neilson, E. and Stinson, G. 2010. Implications of future disturbance regimes on the carbon balance of Canada's managed forest (2010-2100). *Tellus B: Chemical and Physical Meteorology* 62(5), 719-28.

Meyer, N., Welp, G. and Amelung, W. 2018. The temperature sensitivity (Q10) of soil respiration: controlling factors and spatial prediction at regional scale based on environmental soil classes. *Global Biogeochemical Cycles* 32, 306-23.

Molon, M., Boyce, J. I. and Arain, M. A. 2017. Quantitative, nondestructive estimates of coarse root biomass in a temperate pine forest using 3-D ground-penetrating radar (GPR). *Journal of Geophysical Research: Biogeosciences* 122(1), 80-102. Available at: https://agupubs.onlinelibrary.wiley.com/doi/abs/10.1002/2016JG003518.

Murty, D., Kirschbaum, M. U. F., McMurtrie, R. E. and McGilvray, H. 2002. Does conversion of forest to agricultural land change soil carbon and nitrogen? A review of the literature. *Global Change Biology* 8(2), 105-23. doi:10.1046/j.1354-1013.2001.00459.x.

Nabuurs, G.-J., Schelhaas, M.-J., Mohren, G. M. J. and Field, C. B. 2003. Temporal evolution of the European forest sector carbon sink from 1950 to 1999. *Global Change Biology* 9(2), 152-60. doi:10.1046/j.1365-2486.2003.00570.x.

Nabuurs, G.-J., Lindner, M., Verkerk, P. J., Gunia, K., Deda, P., Michalak, R. and Grassi, G. 2013. First signs of carbon sink saturation in European forest biomass. *Nature Climate Change* 3(9), 792-6. doi:10.1038/nclimate1853.

Naipal, V., Ciais, P., Wang, Y., Lauerwald, R., Guenet, B. and Van Oost, K. 2018. Global soil organic carbon removal by water erosion under climate change and land use change during ad 1850-2005. *Biogeosciences* 15(14), 4459-80. doi:10.5194/bg-15-4459-2018.

Naudts, K., Ryder, J., McGrath, M. J., Otto, J., Chen, Y., Valade, A., Bellasen, V., Berhongaray, G., Bönisch, G., Campioli, M., Ghattas, J., De Groote, T., Haverd, V., Kattge, J., MacBean, N., Maignan, F., Merilä, P., Penuelas, J., Peylin, P., Pinty, B., Pretzsch, H., Schulze, E. D., Solyga, D., Vuichard, N., Yan, Y. and Luyssaert, S. 2015. A vertically discretised canopy description for ORCHIDEE (svn r2290) and the modifications to the energy, water and carbon fluxes. *Geoscientific Model Development* 8(7), 2035-65. doi:10.5194/gmd-8-2035-2015.

Naudts, K., Chen, Y., McGrath, M. J., Ryder, J., Valade, A., Otto, J. and Luyssaert, S. 2016. Europe's forest management did not mitigate climate warming. *Science* 351(6273), 597-600. doi:10.1126/science.aad7270.

Nilsson, M. 1996. Estimation of tree heights and stand volume using an airborne LiDAR system. *Remote Sensing of Environment* 56(1), 1-7. doi:10.1016/0034-4257(95)00224-3.

Noormets, A., Epron, D., Domec, J. C., McNulty, S. G., Fox, T., Sun, G. and King, J. S. 2015. Effects of forest management on productivity and carbon sequestration: a review and hypothesis. *Forest Ecology and Management* 355, 124-40. doi:10.1016/j.foreco.2015.05.019.

Norby, R. J., Warren, J. M., Iversen, C. M., Medlyn, B. E. and McMurtrie, R. E. 2010. CO_2 enhancement of forest productivity constrained by limited nitrogen availability.

Proceedings of the National Academy of Sciences of the United States of America 107(45), 19368–73. doi:10.1073/pnas.1006463107.

Oleson, K. W., Lawrence, D. M., Bonan, G. B., Flanner, M. G., Kluzek, E., Lawrence, P. J., Levis, S., Swenson, S. C. and Thornton, P. E. 2010. *Technical Description of Version 4.0 of the Community Land Model (CLM)*. National Center for Atmospheric Research, Boulder, CO.

Olofsson, P., Foody, G. M., Herold, M., Stehman, S. V., Woodcock, C. E. and Wulder, M. A. 2014. Good practices for estimating area and assessing accuracy of land change. *Remote Sensing of Environment* 148, 42–57. doi:10.1016/j.rse.2014.02.015.

Ostonen, I., Truu, M., Helmisaari, H. S., Lukac, M., Borken, W., Vanguelova, E., Godbold, D. L., Löhmus, K., Zang, U., Tedersoo, L., Preem, J. K., Rosenvald, K., Aosaar, J., Armolaitis, K., Frey, J., Kabral, N., Kukumgi, M., Leppälammi-Kujansuu, J., Lindroos, A. J., Meril, P., Napa, Ü., Nöjd, P., Parts, K., Uri, V., Varik, M. and Truu, J. 2017. Adaptive root foraging strategies along a boreal–temperate forest gradient. *The New Phytologist* 215(3), 977–91. doi:10.1111/nph.14643.

Pacala, S. W., Hurtt, G. C., Baker, D., Peylin, P., Houghton, R. A., Birdsey, R. A., Heath, L., Sundquist, E. T., Stallard, R. F., Ciais, P., Moorcroft, P., Caspersen, J. P., Shevliakova, E., Moore, B., Kohlmaier, G., Holland, E., Gloor, M., Harmon, M. E., Fan, S. M., Sarmiento, J. L., Goodale, C. L., Schimel, D. and Field, C. B. 2001. Consistent land- and atmosphere-based U.S. carbon sink estimates. *Science* 292(5525), 2316–20. doi:10.1126/science.1057320.

Pan, Y., Birdsey, R. A., Fang, J., Houghton, R., Kauppi, P. E., Kurz, W. A., Phillips, O. L., Shvidenko, A., Lewis, S. L., Canadell, J. G., Ciais, P., Jackson, R. B., Pacala, S. W., McGuire, A. D., Piao, S., Rautiainen, A., Sitch, S. and Hayes, D. 2011a. A large and persistent carbon sink in the world's forests. *Science* 333(6045), 988–93. doi:10.1126/science.1201609.

Pan, Y., Chen, J. M., Birdsey, R., McCullough, K., He, L. and Deng, F. 2011b. Age structure and disturbance legacy of North American forests. *Biogeosciences* 8(3), 715–32. doi:10.5194/bg-8-715-2011.

Peltoniemi, M., Markkanen, T., Härkönen, S., Muukkonen, P., Thum, T., Aalto, T. and Mäkelä, A. 2015. Consistent estimates of gross primary production of Finnish forests – comparison of estimates of two process models. *Boreal Environment Research* 20, 196–212.

Peng, C., Zhou, X., Zhao, S., Wang, X., Zhu, B., Piao, S. and Fang, J. 2009. Quantifying the response of forest carbon balance to future climate change in northeastern China: model validation and prediction. *Global and Planetary Change* 66(3–4), 179–94. doi:10.1016/j.gloplacha.2008.12.001.

Perugini, L., Caporaso, L., Marconi, S., Cescatti, A., Quesada, B., de Noblet-Ducoudré, N., House, J. I. and Arneth, A. 2017. Biophysical effects on temperature and precipitation due to land cover change. *Environmental Research Letters* 12(5).

Peters, E. B., Wythers, K. R., Zhang, S., Bradford, J. B. and Reich, P. B. 2013. Potential climate change impacts on temperate forest ecosystem processes. *Canadian Journal of Forest Research* 43(10), 939–50. doi:10.1139/cjfr-2013-0013.

Piao, S., Ciais, P., Friedlingstein, P., Peylin, P., Reichstein, M., Luyssaert, S., Margolis, H., Fang, J., Barr, A., Chen, A., Grelle, A., Hollinger, D. Y., Laurila, T., Lindroth, A., Richardson, A. D. and Vesala, T. 2008. Net carbon dioxide losses of northern ecosystems in response to autumn warming. *Nature* 451(7174), 49–52. doi:10.1038/nature06444.

Pielke, R. A., Marland, G., Betts, R. A., Chase, T. N., Eastman, J. L., Niles, J. O., Niyogi, D. D. and Running, S. W. 2002. The influence of land-use change and landscape dynamics on the climate system: relevance to climate-change policy beyond the radiative effect of greenhouse gases. *Philosophical Transactions. Series A, Mathematical, Physical, and Engineering Sciences* 360(1797), 1705–19. doi:10.1098/rsta.2002.1027.

Pilli, R., Grassi, G., Kurz, W. A., Fiorese, G. and Cescatti, A. 2017. The European forest sector: past and future carbon budget and fluxes under different management scenarios. *Biogeosciences* 14(9), 2387–405. doi:10.5194/bg-14-2387-2017.

Pongratz, J., Reick, C. H., Raddatz, T., Caldeira, K. and Claussen, M. 2011. Past land use decisions have increased mitigation potential of reforestation. *Geophysical Research Letters* 38(15), L15701. doi:10.1029/2011GL047848.

Pongratz, J., Dolman, H., Don, A., Erb, K. H., Fuchs, R., Herold, M., Jones, C., Kuemmerle, T., Luyssaert, S., Meyfroidt, P. and Naudts, K. 2018. Models meet data: challenges and opportunities in implementing land management in earth system models. *Global Change Biology* 24(4), 1470–87. doi:10.1111/gcb.13988.

Popescu, S. C., Wynne, R. H. and Nelson, R. F. 2003. Measuring individual tree crown diameter with LiDAR and assessing its influence on estimating forest volume and biomass. *Canadian Journal of Remote Sensing* 29(5), 564–77. doi:10.5589/m03-027.

Pregitzer, K. and Euskirchen, E. 2004. Carbon cycling and storage in world forests: biome patterns related to forest age. *Global Change Biology* 10(12), 2052–77.

Prestele, R., Alexander, P., Rounsevell, M. D. A., Arneth, A., Calvin, K., Doelman, J., Eitelberg, D. A., Engstrm, K., Fujimori, S., Hasegawa, T., Havlik, P., Humpender, F., Jain, A. K., Krisztin, T., Kyle, P., Meiyappan, P., Popp, A., Sands, R. D., Schaldach, R., Schngel, J., Stehfest, E., Tabeau, A., Van Meijl, H., Van Vliet, J. and Verburg, P. H. 2016. Hotspots of uncertainty in land-use and land-cover change projections: a global-scale model comparison. *Global Change Biology* 22(12), 3967–83. doi:10.1111/gcb.13337.

Pretzsch, H., Biber, P., Schütze, G., Uhl, E. and Rötzer, T. 2014. Forest stand growth dynamics in Central Europe have accelerated since 1870. *Nature Communications* 5, 4967. doi:10.1038/ncomms5967.

Pugh, T. A. M., Jones, C. D., Huntingford, C., Burton, C., Lomas, M., Robertson, E., Piao, S. L. and Sitch, S. 2018. A large committed long-term sink of carbon due to vegetation dynamics. *Earth's Future* 6(1), 1–20. doi:10.1002/eft2.238.

Puhlick, J. J., Weiskittel, A. R., Fernandez, I. J., Fraver, S., Kenefic, L. S., Seymour, R. S., Kolka, R. K., Rustad, L. E. and Brissette, J. C. 2016. Long-term influence of alternative forest management treatments on total ecosystem and wood product carbon storage. *Canadian Journal of Forest Research* 46(11), 1404–12. doi:10.1139/cjfr-2016-0193.

Rabin, S. S., Melton, J. R., Lasslop, G., Bachelet, D., Forrest, M., Hantson, S., Li, F., Mangeon, S., Yue, C., Arora, V. K., Hickler, T., Kloster, S., Knorr, W., Nieradzik, L., Spessa, A., Folberth, G. A., Sheehan, T., Voulgarakis, A., Kelley, D. I., Prentice, I. C., Sitch, S., Harrison, S. and Arneth, A. 2017. The fire modeling intercomparison project (FireMIP), phase 1: experimental and analytical protocols. *Geoscientific Model Development* 20, 1175–97.

Reick, C. H., Raddatz, T., Pongratz, J. and Claussen, M. 2010. Contribution of anthropogenic land cover change emissions to pre-industrial atmospheric CO_2. *Tellus B: Chemical and Physical Meteorology* 62(5), 329–36. doi:10.1111/j.1600-0889.2010.00479.x.

Ryan, M. G., Binkley, D. and Fownes, J. H. 1997. Age-related decline in forest productivity: pattern and process. *Advances in Ecological Research* 27, 213–62.

Saban, J. M., Chapman, M. A. and Taylor, G. 2019. FACE facts hold for multiple generations; evidence from natural CO_2 springs. *Global Change Biology* 25(1), 1–11. doi:10.1111/gcb.14437.

Sadeghi, Y., St-Onge, B., Leblon, B., Prieur, J.-F. and Simard, M. 2018. Mapping boreal forest biomass from a SRTM and TanDEM-X based on canopy height model and Landsat spectral indices. *International Journal of Applied Earth Observation and Geoinformation* 68, 202–13. doi:10.1016/j.jag.2017.12.004.

Saeki, T., Maksyutov, S., Sasakawa, M., Machida, T., Arshinov, M., Tans, P., Conway, T. J., Saito, M., Valsala, V., Oda, T., Andres, R. J. and Belikov, D. 2013. Carbon flux estimation for Siberia by inverse modeling constrained by aircraft and tower CO_2 measurements. *Journal of Geophysical Research: Atmospheres* 118(2), 1100–22.

Schelhaas, M. J., van Esch, P. W., Groen, T. A., de Jong, B. H. J., Kanninen, M., Liski, J., Masera, O., Mohren, G. M. J., Nabuurs, G. J., Palosuo, T., Pedroni, L., Vallejo, A. and Vilén, T. 2004. *CO2FIX V 3.1 – A Modelling Framework for Quantifying Carbon Sequestration in Forest Ecosystems*. ALTERRA, Wageningen, the Netherlands.

Schelhaas, M. J., Eggers, J., Lindner, M., Nabuurs, G. J., Pussinen, A., Paivinen, R., Schuck, A., Verkerk, P. J., van der Werf, D. C. and Zudin, S. 2007. *Model Documentation for the European Forest Information Scenario Model (EFISCEN 3.1.3)*.

Schelhaas, M.-J., Hengeveld, G. M., Heidema, N., Thürig, E., Rohner, B., Vacchiano, G., Vayreda, J., Redmond, J., Socha, J., Fridman, J., Tomter, S., Polley, H., Barreiro, S. and Nabuurs, G.-J. 2018. Species-specific, pan-European diameter increment models based on data of 2.3 million trees. *Forest Ecosystems* 5(1), 21. doi:10.1186/s40663-018-0133-3.

Schimel, D. S., House, J. I., Hibbard, K. A., Bousquet, P., Ciais, P., Peylin, P., Braswell, B. H., Apps, M. J., Baker, D., Bondeau, A., Canadell, J., Churkina, G., Cramer, W., Denning, A. S., Field, C. B., Friedlingstein, P., Goodale, C., Heimann, M., Houghton, R. A., Melillo, J. M., Moore, B., Murdiyarso, D., Noble, I., Pacala, S. W., Prentice, I. C., Raupach, M. R., Rayner, P. J., Scholes, R. J., Steffen, W. L. and Wirth, C. 2001. Recent patterns and mechanisms of carbon exchange by terrestrial ecosystems. *Nature* 414(6860), 169–72. doi:10.1038/35102500.

Schindlbacher, A., Borken, W., Djukic, I., Brandstätter, C., Spötl, C. and Wanek, W. 2015a. Contribution of carbonate weathering to the CO_2 efflux from temperate forest soils. *Biogeochemistry* 124(1–3), 273–90. doi:10.1007/s10533-015-0097-0.

Schindlbacher, A., Schnecker, J., Takriti, M., Borken, W. and Wanek, W. 2015b. Microbial physiology and soil CO_2 efflux after 9 years of soil warming in a temperate forest – no indications for thermal adaptations. *Global Change Biology* 21(11), 4265–77. doi:10.1111/gcb.12996.

Schroeder, L. M. 2010. Colonization of storm gaps by the spruce bark beetle: influence of gap and landscape characteristics. *Agricultural and Forest Entomology* 12(1), 29–39.

Schulze, E. D., Luyssaert, S., Ciais, P., Freibauer, A., Janssens, I. A., Soussana, J. F., Smith, P., Grace, J., Levin, I., Thiruchittampalam, B., Heimann, M., Dolman, A. J., Valentini, R., Bousquet, P., Peylin, P., Peters, W., Rödenbeck, C., Etiope, G., Vuichard, N., Wattenbach, M., Nabuurs, G. J., Poussi, Z., Nieschulze, J., Gash, J. H. and the CarboEurope Team 2009. Importance of methane and nitrous oxide for Europe's terrestrial greenhouse-gas balance. *Nature Geoscience* 2, 842–50.

Seidl, R., Schelhaas, M. J., Rammer, W. and Verkerk, P. J. 2014. Increasing forest disturbances in Europe and their impact on carbon storage. *Nature Climate Change* 4(9), 806–10. doi:10.1038/nclimate2318.

Shabaga, J. A., Basiliko, N., Caspersen, J. P. and Jones, T. A. 2017. Skid trail use influences soil carbon flux and nutrient pools in a temperate hardwood forest. *Forest Ecology and Management* 402, 51–62.

Shao, W., Cai, J., Wu, H., Liu, J., Zhang, H. and Huang, H. 2017. An assessment of carbon storage in chinas arboreal forests. *Forests* 8, 110.

Shukla, J., Nobre, C. and Sellers, P. 1990. Amazon deforestation and climate change. *Science* 247(4948), 1322–5. doi:10.1126/science.247.4948.1322.

Sillett, S. C., Pelt, R. V., Kramer, R. D., Carroll, A. L. and Koch, G. W. 2015. Biomass and growth potential of Eucalyptus regnans up to 100m tall. *Forest Ecology and Management* 348, 78 – 91. doi:10.1016/j.foreco.2015.03.046.

Smith, P., House, J. I., Bustamante, M., Sobocka, J., Harper, R., Pan, G., West, P. C., Clark, J. M., Adhya, T., Rumpel, C., Paustian, K., Kuikman, P., Cotrufo, M. F., Elliott, J. A., McDowell, R., Griffiths, R. I., Asakawa, S., Bondeau, A., Jain, A. K., Meersmans, J. and Pugh, T. A. M. 2016. Global change pressures on soils from land use and management. *Global Change Biology* 22(3), 1008–28. doi:10.1111/gcb.13068.

Solly, E. F., Brunner, I., Helmisaari, H. S., Herzog, C., Leppälammi-Kujansuu, J., Schöning, I., Schrumpf, M., Schweingruber, F. H., Trumbore, S. E. and Hagedorn, F. 2018. Unravelling the age of fine roots of temperate and boreal forests. *Nature Communications* 9(1), 3006. doi:10.1038/s41467-018-05460-6.

Steinkamp, K., Fletcher, S. E. M., Brailsford, G., Smale, D., Moore, S., Keller, E. D., Baisden, W. T., Mukai, H. and Stephens, B. B. 2017. Atmospheric CO_2 observations and models suggest strong carbon uptake by forests in New Zealand. *Atmospheric Chemistry and Physics* 17, 47–76.

Stendahl, J., Berg, B. and Lindahl, B. D. 2017. Manganese availability is negatively associated with carbon storage in northern coniferous forest humus layers. *Scientific Reports* 7(1), 15487. doi:10.1038/s41598-017-15801-y.

Stone, E., Harris, W., Brown, R. and Kuehl, R. 1993. Carbon storage in Florida spodosols. *Soil Science Society of America Journal* 57, 179–82.

Stovall, A. E., Vorster, A. G., Anderson, R. S., Evangelista, P. H. and Shugart, H. H. 2017. Non-destructive aboveground biomass estimation of coniferous trees using terrestrial LiDAR. *Remote Sensing of Environment* 200, 31–42.

Strickland, M. S., Devore, J. L., Maerz, J. C. and Bradford, M. A. 2010. Grass invasion of a hardwood forest is associated with declines in belowground carbon pools. *Global Change Biology* 16(4), 1338–50.

Sulkava, M., Luyssaert, S., Zaehle, S. and Papale, D. 2011. Assessing and improving the representativeness of monitoring networks: the European flux tower network example. *Journal of Geophysical Research: Biogeosciences* 116(G3), G00J04. doi:10.1029/2010JG001562.

Sullivan, M. J., Talbot, J., Lewis, S. L., Phillips, O. L., Qie, L., Begne, S. K., Chave, J., Cuni-Sanchez, A., Hubau, W., Lopez-Gonzalez, G., Miles, L., Monteagudo-Mendoza, A., Sonké, B., Sunderland, T., Ter Steege, H., White, L. J., Affum-Baffoe, K., Aiba, S. I., de Almeida, E. C., de Oliveira, E. A., Alvarez-Loayza, P., Dávila, E. Á, Andrade, A., Aragão, L. E., Ashton, P., Aymard C, G. A., Baker, T. R., Balinga, M., Banin, L. F., Baraloto, C., Bastin, J. F., Berry, N., Bogaert, J., Bonal, D., Bongers, F., Brienen, R., Camargo, J. L., Cerón, C., Moscoso, V. C., Chezeaux, E., Clark, C. J., Pacheco, Á. C., Comiskey, J. A., Valverde, F. C., Coronado, E. N., Dargie, G., Davies, S. J., De Canniere, C., Djuikouo K, M. N., Doucet, J. L., Erwin, T. L., Espejo, J. S., Ewango, C. E., Fauset, S., Feldpausch, T. R., Herrera, R., Gilpin, M., Gloor, E., Hall, J. S., Harris, D. J., Hart, T. B., Kartawinata, K.,

Kho, L. K., Kitayama, K., Laurance, S. G., Laurance, W. F., Leal, M. E., Lovejoy, T., Lovett, J. C., Lukasu, F. M., Makana, J. R., Malhi, Y., Maracahipes, L., Marimon, B. S., Junior, B. H., Marshall, A. R., Morandi, P. S., Mukendi, J. T., Mukinzi, J., Nilus, R., Vargas, P. N., Camacho, N. C., Pardo, G., Peña-Claros, M., Pétronelli, P., Pickavance, G. C., Poulsen, A. D., Poulsen, J. R., Primack, R. B., Priyadi, H., Quesada, C. A., Reitsma, J., Réjou-Méchain, M., Restrepo, Z., Rutishauser, E., Salim, K. A., Salomão, R. P., Samsoedin, I., Sheil, D., Sierra, R., Silveira, M., Slik, J. W., Steel, L., Taedoumg, H., Tan, S., Terborgh, J. W., Thomas, S. C., Toledo, M., Umunay, P. M., Gamarra, L. V., Vieira, I. C., Vos, V. A., Wang, O., Willcock, S. and Zemagho, L. 2017. Diversity and carbon storage across the tropical forest biome. *Scientific Reports* 7, 39102. doi:10.1038/srep39102.

Taylor, A. R., Seedre, M., Brassard, B. W. and Chen, H. Y. H. 2014. Decline in net ecosystem productivity following canopy transition to late-succession forests. *Ecosystems* 17(5), 778–91. doi:10.1007/s10021-014-9759-3.

Thom, D. and Seidl, R. 2016. Natural disturbance impacts on ecosystem services and biodiversity in temperate and boreal forests. *Biological Reviews of the Cambridge Philosophical Society* 91(3), 760–81. doi:10.1111/brv.12193.

Thompson, R. L. and Stohl, A. 2014. Flexinvert: an atmospheric bayesian inversion framework for determining surface fluxes of trace species using an optimized grid. *Geoscientific Model Development* 7(5), 2223–42. doi:10.5194/gmd-7-2223-2014.

Trettin, C. C. and Jurgensen, M. F. 2003. *Carbon Cycling in Wetland Forest Soils*. Lewis Publishers, Boca Raton, London, New York and Washington DC.

Tropek, R., Sedláček, O., Beck, J., Keil, P., Musilová, Z., Šímová, I. and Storch, D. 2014. Comment on 'high-resolution global maps of 21st-century forest cover change'. *Science* 344(6187), 981–. doi:10.1126/science.1248753.

Tunved, P., Hansson, H. C., Kerminen, V. M., Ström, J., Maso, M. D., Lihavainen, H., Viisanen, Y., Aalto, P. P., Komppula, M. and Kulmala, M. 2006. High natural aerosol loading over boreal forests. *Science* 312(5771), 261–3. doi:10.1126/science.1123052.

Tuomi, M., Thum, T., Jrvinen, H., Fronzek, S., Berg, B., Harmon, M., Trofymow, J., Sevanto, S. and Liski, J. 2009. Leaf litter decomposition estimates of global variability based on YASSO07 model. *Ecological Modelling* 220(23), 3362–71.

Uglietti, C., Leuenberger, M. C. and Brunner, D. 2011. European source and sink areas of CO_2 retrieved from Lagrangian transport model interpretation of combined CO_2 and CO_2 measurements at the high alpine research station Jungfraujoch. *Atmospheric Chemistry and Physics* 11(15), 8017–36. doi:10.5194/acp-11-8017-2011.

UNFCCC. n.d. *NDC Registry (Interim)*. Available at: http://www4.unfccc.int/ndcregistry/Pages/All.aspx (accessed on November 06, 2018).

Ung, C.-H., Bernier, P. and Guoa, X.-J. 2008. Canadian national biomass equations: new parameter estimates that include British Columbia data. *Canadian Journal of Forest Research* 38, 1123–32.

Urbano, A. R. and Keeton, W. S. 2017. Carbon dynamics and structural development in recovering secondary forests of the northeastern U.S. *Forest Ecology and Management* 392, 21–35. doi:10.1016/j.foreco.2017.02.037.

Valade, A., Bellassen, V., Magand, C. and Luyssaert, S. 2017. Sustaining the sequestration efficiency of the European forest sector. *Forest Ecology and Management* 405, 44–55. doi:10.1016/j.foreco.2017.09.009.

Van Oijen, M., Ågren, G. I., Chertov, O., Kahle, H.-P., Kellomäki, S., Komarov, A., Mellert, K., Prietzel, J., Spiecker, H. and Straussberger, R. 2008. 'A comparison of empirical and process-based modelling methods for analysing changes in European forest growth', causes and consequences of forest growth trends in Europe, European Forest Institute. *Research Report* 21, 203–16.

Vastaranta, M., Yu, X., Luoma, V., Karjalainen, M., Saarinen, N., Wulder, M. A., White, J. C., Persson, H. J., Hollaus, M., Yrttimaa, T., Holopainen, M. and Hyyppä, J. 2018. Aboveground forest biomass derived using multiple dates of WorldView-2 stereo-imagery: quantifying the improvement in estimation accuracy. *International Journal of Remote Sensing* 39(23), 8766–83.

Verkerk, P. J., Schelhaas, M. J., Immonen, V., Hengeveld, G., Kiljunen, J., Lindner, M., Nabuurs, G. J., Suominen, T. and Zudin, S. 2016. *Manual for the European Forest Information Scenario Model (EFISCEN 4.1)*. European Forest Institute, Joensuu, Finland.

Vieilledent, G., Fischer, F. J., Chave, J., Guibal, D., Langbour, P. and Gérard, J. 2018. New formula and conversion factor to compute basic wood density of tree species using a global wood technology database. *American Journal of Botany* 105(10), 1653–61. doi:10.1002/ajb2.1175.

Vitousek, P. M. and Howarth, R. W. 1991. Nitrogen limitation on land and in the sea: how can it occur? *Biogeochemistry* 13(2), 87–115.

Vuichard, N., Messina, P., Luyssaert, S., Guenet, B., Zaehle, S., Ghattas, J., Bastrikov, V. and Peylin, P. 2018. Accounting for carbon and nitrogen interactions in the global terrestrial ecosystem model ORCHIDEE (trunk version, rev 4999): multi-scale evaluation of gross primary production. *Geoscientific Model Development Discussions* 2018, 1–41. doi:10.5194/gmd-2018-261.

Wang, C. 2006. Biomass allometric equations for 10 co-occurring tree species in Chinese temperate forests. *Forest Ecology and Management* 222(1), 9–16.

Wear, D. N. and Coulston, J. W. 2015. From sink to source: regional variation in us forest carbon futures. *Scientific Reports* 5, 16518. doi:10.1038/srep16518.

Woodwell, G. M., Hobbie, J. E., Houghton, R. A., Melillo, J. M., Moore, B., Peterson, B. J. and Shaver, G. R. 1983. Global deforestation: contribution to atmospheric carbon dioxide. *Science* 222(4628), 1081–6. doi:10.1126/science.222.4628.1081.

Yang, X., Tang, J., Mustard, J. F., Lee, J.-E., Rossini, M., Joiner, J., Munger, J. W., Kornfeld, A. and Richardson, A. D. 2015. Solar-induced chlorophyll fluorescence that correlates with canopy photosynthesis on diurnal and seasonal scales in a temperate deciduous forest. *Geophysical Research Letters* 42(8), 2977–87. doi:10.1002/2015GL063201.

Yuan, Z., Wang, S., Ali, A., Gazol, A., Ruiz-Benito, P., Wang, X., Lin, F., Ye, J., Hao, Z. and Loreau, M. 2018. Aboveground carbon storage is driven by functional trait composition and stand structural attributes rather than biodiversity in temperate mixed forests recovering from disturbances. *Annals of Forest Science* 75, 67.

Zaehle, S., Medlyn, B. E., De Kauwe, M. G., Walker, A. P., Dietze, M. C., Hickler, T., Luo, Y., Wang, Y. P., El-Masri, B., Thornton, P., Jain, A., Wang, S., Warlind, D., Weng, E., Parton, W., Iversen, C. M., Gallet-Budynek, A., McCarthy, H., Finzi, A. C., Hanson, P. J., Prentice, I. C., Oren, R. and Norby, R. J. 2014. Evaluation of 11 terrestrial carbon-nitrogen cycle models against observations from two temperate Free-Air CO_2 Enrichment studies. *The New Phytologist* 202(3), 803–22. doi:10.1111/nph.12697.

Zhang, C., Ju, W., Chen, J. M., Zan, M., Li, D., Zhou, Y. and Wang, X. 2013. China's forest biomass carbon sink based on seven inventories from 1973 to 2008. *Climatic Change* 118(3–4), 933–48. doi:10.1007/s10584-012-0666-3.

Zhang, G., Ganguly, S., Nemani, R. R., White, M. A., Milesi, C., Hashimoto, H., Wang, W., Saatchi, S., Yu, Y. and Myneni, R. B. 2014. Estimation of forest aboveground biomass in California using canopy height and leaf area index estimated from satellite data. *Remote Sensing of Environment* 151, 44–56.

Ziche, D., Grüneberg, E., Hilbrig, L., Höhle, J., Kompa, T., Liski, J., Repo, A. and Wellbrock, N. 2019. Comparing soil inventory with modelling: carbon balance in central European forest soils varies among forest types. *The Science of the Total Environment* 647, 1573–85. doi:10.1016/j.scitotenv.2018.07.327.

Chapter 5

Climate change and tropical forests

Rodney J. Keenan, The University of Melbourne, Australia

1 Introduction

Climate change caused by increased greenhouse gas (GHG) emissions is a major global challenge. Tropical forests are central to addressing this challenge. In recent decades, the tropics have contributed by far the greatest proportion of GHG emissions due to deforestation and forest degradation (Liu et al., 2015). Fifty-two percent of global forests are found in the tropical and subtropical zones, and tropical forests provide the habitat for much of the world's biodiversity. The tropics are also home to 40% of the world's people, and tropical countries have the most rapid population growth with 50% of the global population expected to be living in the tropics by 2030. The tropics contain most of the world's poorest countries, and this has large implications for both potential forest loss and the vulnerability and potential future impacts of climate change on forests and people.

An expanding tropical population will place increased demands on the ecosystem goods and services provided from tropical forests. Over the course of human history, most deforestation happened in the temperate regions of East Asia, Europe, North America, and South America. Since the 1950s, forested areas in temperate regions have increased, and deforestation now mainly occurs

http://dx.doi.org/10.19103/AS.2020.0074.09

in the tropics (Keenan et al., 2015). Slowing deforestation and degradation, and more sustainable management of tropical forests, can reduce GHG emissions and the rate of future climate change. Restoring and sustainably managing forests can increase the resilience of ecosystems, support adaptation to climate change in tropical societies, and increase provision of other environmental services. To achieve these outcomes, there will need to be an increasing focus on adaptation in tropical forest management (Seppala et al., 2009).

Given the importance of tropical forests in both mitigating and adapting to climate change, they have become central in international climate change agreements. Forest-based mitigation options were included in the first round of emission reduction commitments for developed countries under the 1997 Kyoto Protocol but investment in tropical forests through Kyoto mechanisms was modest. A more recent focus in the UNFCCC on REDD+ has provided a platform for a greater focus on tropical forests. The New York Declaration on Forests (UN, 2014) promoted including land use and land use change in the 2015 climate agreement and committed signatories to reducing the rate of loss of natural forests and major forest restoration targets. The FAO indicates in its 2050 vision on forests and forestry that forests are an essential solution to climate change adaptation and mitigation. The Paris Agreement reinforces the role of forests more generally in climate change action. Signatories to the Bonn Challenge have committed to restoring 150 million ha of degraded landscapes and forestlands by 2020 and an additional 200 million ha by 2030 (Stanturf et al., 2019).

This chapter provides an overview of the interactions between tropical forests and climate. It begins by presenting the role of tropical forests in the global carbon cycle and the relationship between forests and regional and local climate; then describes the current and potential future impacts of climate change on forests in different parts of the tropics; and presents an analysis of the options for future management of forests to adapt to climate change. The chapter concludes by identifying knowledge gaps and areas for future research.

2 Tropical forests and the global carbon cycle

Forests are a major carbon (C) sink worldwide (Pan et al., 2011). Tropical forests are among the most C dense ecosystems, storing up to 55% of the planet's total aboveground C. Tropical forests are dynamic, with the total area at any time determined by the loss of forests in some places and the expansion of forests in other places. Regions also can undergo forest transitions, changing over time from net forest loss to net forest gain (or vice versa). Tropical deforestation is a significant contributor to global warming and climate change. It results in the release of carbon stored in wood as carbon dioxide into the atmosphere. Deforestation and forest degradation accounted for about a third of anthropogenic CO_2 emissions from 1750 to 2011 and 12% of annual emissions between 2000 and 2009 (IPCC, 2013). Since the 1950s, deforestation emissions have largely occurred in the tropics. Present-day assessments indicate that the global rate of deforestation has declined, with net forest loss in the tropics decreasing from 10.4 million ha per year in the 1990s to 6.4 million ha per year over the period from 2010 to 2015. However, deforestation rates remain high in Latin America, Southeast Asia, and Africa (Keenan et al., 2015).

Globally, land is a net sink for CO_2, although this fluctuates from year to year (Le Quéré et al., 2015). Modeling suggests this sink is a result of the recovery from past disturbance, the indirect effects of human activity, such as increased levels of CO_2 and N in the atmosphere, and a warming climate in temperate regions (Erb et al., 2013). This recovery is focused more in temperate regions. Considering the effects of deforestation, degradation, and the increase in carbon stock in recovering forests, the contribution of tropical forests to the global carbon cycle is currently about neutral, with intact and recovering forests taking up as much carbon as is released through deforestation and degradation (Mitchard, 2018). Rising atmospheric CO_2 might increase tropical forest carbon uptake, but these are likely to be offset by the negative impacts of climate change (see in the following paragraph).

The underlying drivers of deforestation include population growth, economic development, and demand for food, fiber, and energy. The primary causes of deforestation are conversion of forests to small- and large-scale agriculture and livestock grazing, unsustainable charcoal production and fuelwood collection, and the development of mines, hydropower dams, and other infrastructure items, such as roads and powerlines (Meyfroidt et al., 2018). Dominant agricultural crops associated with deforestation are oil palm, soybeans, and sugar. Unsustainable and uncontrolled timber harvesting is also an important secondary cause of deforestation because it opens access for deforestation for small- and large-scale agriculture. Deforestation is also high in less-industrialized countries with high population-growth rates, particularly in Africa, due to the actions of small-scale farmers (Keenan et al., 2015).

Gross emissions due to temporary forest loss through harvesting can be as high as for permanent deforestation, but are largely balanced by uptake in regrowing forest, so net emissions are small. Degradation was responsible for 15% of total net emissions from tropical forests (Baccini et al., 2012) and in some tropical regions, regrowth may not balance removals, leading to degradation of forest carbon stocks. For example, forest degradation in the Amazon, mostly due to selective logging, was considered responsible for 15–19% higher C emissions than those from deforestation alone (Huang and Asner, 2010). Shifting (or swidden) cultivation also impacts forest carbon stocks. Swidden is a traditional form of agriculture that involves temporary clearing and burning of forests and growing agricultural crops for household consumption or sale. After

3-5 years of crop use, the land is left alone for 5-20 years, forests regenerate, and soil fertility is restored; then, the cycle is repeated. Carbon stocks in areas subject to long-term cycles of farming and fallow are likely to be stable, but expansion of swidden into previously unfarmed forests, or shortening of the fallow cycle, are likely to decrease landscape carbon stocks (Dressler et al., 2015).

Without effective actions to reduce deforestation, tropical forests are likely to become a carbon source in the future, due to continued forest loss and the impacts of climate change on carbon sequestration capacity. This could occur through increasing frequency or intensity of the El Nino South Oscillation phenomena. For example, Liu et al. (2017) found that in response to the warmer and drier climate associated with the 2014-2015 El Nino, the pantropical biosphere released 2.5 ± 0.34 gigatons more carbon into the atmosphere than released during 2011. This contributed to the highest increase in atmospheric CO_2 ever recorded in 2015. The three tropical continents all had higher forest emissions, but the dominant processes varied. Gross primary production (GPP) reduced carbon uptake in tropical South America, fire increased carbon release in tropical Asia, and respiration increased carbon release in Africa. Higher carbon release was associated with either extremely low precipitation or high temperatures, or both.

3 Other climate effects of tropical forests

Forest conversion increases surface albedo (reflectance), changes in evapotranspiration, and surface roughness. These affect latent heat flux and the hydrological cycle. The effects of these on local climate will differ depending on regional climate regime and forest types. For example, loss of tropical forests increases warming potential through lower evaporative cooling (Bonan, 2008).

Recent analysis (Alkama and Cescatti, 2016) indicates that in all climate zones, forest clearing produces a marked increase of mean annual maximum air surface temperatures, slight changes in minimum temperatures, and an overall increase in mean temperatures, except at the northernmost latitudes. The removal of forest cover in the temperate and tropical zones increases the temperature by about 1°C and by more than 2°C in the arid zone. This daytime warming leads to an increase in the diurnal variation (the difference between the daily maximum and minimum temperature) in deforested areas of about $1.95 \pm 0.08°C$ in tropical climate zones, respectively. Sensitivity of land surface temperature to changes in forest cover is about 50% larger than that of air temperature. The key role of forest evapotranspiration is illustrated through the strongest effects being exhibited in arid areas. Tropical forests, therefore, have an important effect on local maximum temperature and decreasing diurnal and annual temperature variations. The assessment did not capture the signal of large-scale, land-atmosphere interactions or regional teleconnections. Afforestation or reforestation can significantly reduce the biophysical effect of forest clearing on surface temperature.

4 Changing climate in the tropics

The Earth's climate varies on multiple timescales due to a range of factors. The evidence is increasingly clear that since the 1950s there is a strong human signal in warming due to rising atmospheric GHG concentrations. Latest estimates from the IPCC indicate that the global mean surface temperature increase will likely be in the range of 0.3-0.7°C for the period 2016-2035 relative to 1986-2005. Tropical mean annual temperatures (MAT) have followed an increasing trend for the past 130 years, with a modeled rate of increase of 0.1°C per decade between 1900 and 1950 (IPCC, 2013) and an observed average increase of 0.26°C per decade since the mid-1970s. MAT in tropical regions reached peaks of 28°C in the 2000s (Corlett, 2011).

Given the lagged nature of climate responses to GHG, changes associated with past emissions are 'locked in' to the climate system for the next 20 years or so, or even longer for some factors such as sea level rise. Beyond this, projections of global climate are dependent on the future trajectory of GHG emissions. Future emission trajectories are summarized into 'representative concentration pathways' that are dependent on the speed of technology development and the uptake of new technologies in energy, transport, or industrial factors (Van Vuuren et al., 2011).

Near-term increases in seasonal mean and annual mean temperatures are expected to be larger in the tropics and subtropics than in the mid-latitudes (Table 1). This means more frequent hot and fewer cold temperature extremes over most land areas on daily and seasonal timescales, and increased frequency and duration of heat waves (IPCC, 2013). In many mid-latitude and subtropical dry regions, mean precipitation will likely decrease, while in many mid-latitude wet regions, mean precipitation will likely increase. Extreme precipitation events over mid-latitude, wet, tropical regions are likely to become more intense and more frequent as global mean surface temperature increases. The area encompassed by monsoon systems will also increase over the twenty-first century. Monsoon winds may weaken, but monsoon precipitation is likely to intensify due to increased atmospheric moisture and monsoons may start earlier and finish later. The precipitation variability of the El Niño-Southern Oscillation (ENSO), which affects many tropical forests through Asia, Oceania, and the Americas, will remain the dominant mode of interannual variability in the tropical Pacific. Due to the increase in moisture availability, ENSO-related variability will likely intensify, although the unpredictability of ENSO means

Table 1 Potential impacts of climate change in tropical forest climatic regions

Forest region	Future climatic changes	Impacts on forests
Tropical moist forests (Am)	Increased atmospheric CO_2 Increased temperature Increased storm and cyclone intensity Increased dry season fire risks Increased flooding Coastal inundation	Potential increased forest productivity on sites with better soils More introduced species (both weed and tropical native species) Loss of animal and plant biodiversity at higher elevations Impacts on ecosystem functioning associated with new species, storm, and increased fire risks. Impacts on coastal ecosystems, mangroves, and wetlands
Tropical savannah (Aw)	Increased atmospheric CO_2 Increased temperature Less rainfall	Potential expansion of forest on sites with better soils Changes in species distribution Exacerbation of forest degradation with continued threats such as land clearing, weed invasions, or feral animals
Tropical monsoon forests (Am)	Increased atmospheric CO_2 Increased temperature Changes in rainfall pattern in the south with longer dry season and more extended drought	Potential increased forest productivity on sites with better soils Drought, fire, and weed potential will increase Loss or change of the character of high-value biodiversity sites
Higher altitude tropical forests	Increased atmospheric CO_2 Increased temperature	Increased plant water stress in summer High risk of loss of high altitude species Increasing thermophilic species

that projections about the extent of change remain uncertain. The intensity of tropical hurricanes is likely to increase and these will track to higher latitudes and impact on larger areas (IPCC, 2013). Drying conditions in some parts of the tropics will increase the risk of wildfire (Jolly et al., 2015).

On a regional scale, climate changes are much harder to predict as some areas might become cooler while others may warm up more than the global average. For the Asia Pacific region, it has been projected that the temperature will increase by about 0.8°C in 2030 and by 2.1°C by 2070 (Preston et al., 2006). Oceans are warming faster than the land and the strongest ocean warming is projected in tropical and Northern Hemisphere subtropical regions. Thermal expansion due to ocean warming and contributions from melting glaciers and ice sheets means that global mean sea level rise for 2081–2100 relative to 1986–2005 will range from 0.26 to 0.98 m, depending on the extent of increase in atmospheric GHGs (IPCC, 2013).

5 Climate change impacts on tropical forests

The impacts of climate change on forests will be realized in many ways: changes in the distribution of species and ecosystems, impacts on forest composition and structure, effects on ecosystem processes and functions, and the capacity of forests to supply different goods and ecosystem services. Research on climate change impacts in tropical forests is not as extensive, or as long term, as in temperate and boreal forests. Clark (2007) identified three major challenges in assessing responses of tropical forests to climate change: (i) our limited understanding of the effects of past climate, natural disturbances or human activities in tropical forest study sites; (ii) the variation in methods used and absence of long-term monitoring data against which climate-driven changes within forests might be detected; and (iii) the complexity of forest responses to environmental change, with a range of changes occurring simultaneously that are integrated in plant and animal responses that have diverse forms of lags and nonlinearities.

Corlett (2012) similarly lamented the lack of hard information on which to consider responses of tropical forests to climate change. Widely used climate projections for the tropics do not fully represent future possibilities or likely future rates of GHG emissions. The 2-3°C rise commonly assumed may be 4-6, or even 7°C, and projections for rainfall and other climate variables have still greater uncertainty. This lack of knowledge is compounded by the limited understanding about the potential biological consequences of these changes.

A further complication in assessing climate impacts on tropical forests is the variety of forest types and climate conditions found within the 'tropics'. Tropical forests can be classified according to climate zone (Fig. 1) or by species composition and functioning (Montagnini and Jordan, 2005). Land in tropical latitudes varies in altitude (and therefore temperature regime), rainfall amount and pattern, and underlying geology and soil types. Tropical forests tend to occur on highly weathered soils, ultisols, and oxisols, with high soil N and low soil P and cation availability, but they also occur on more recent, less weathered soils with higher P and cation availability (Cusack et al., 2016). High mountain ranges in the tropics, such as the Himalayas, the Andes, and the mountain peaks in Africa, are effectively 'third poles' with climate conditions at altitude similar to lowland areas at much higher latitudes (https://globalforestatlas.yale.edu/tropical-zone/himalayan-forest). The species composition and ecosystem functioning in these higher altitude zones are, therefore, more like temperate and boreal forests. These vegetation types are often tightly stratified in narrow altitudinal zones. Therefore, these mountain forests are more at risk from climate change because the species in these zones need to move more quickly to stay within their ecological and climatic

Tropical (1980–2016)

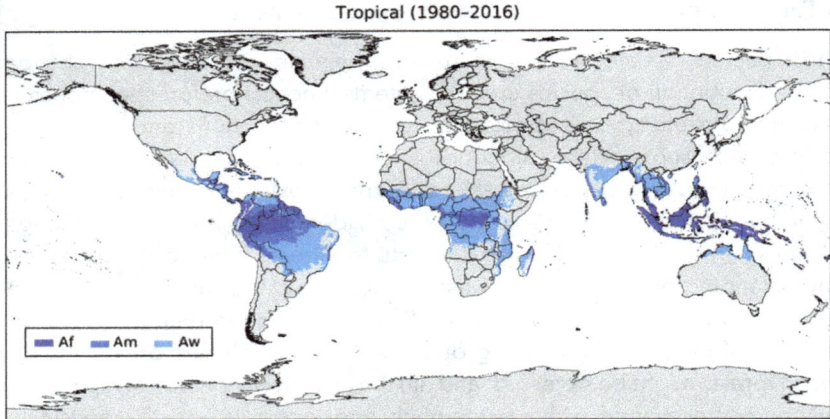

Figure 1 Tropical forest climatic zones. Af – Rainforest, Am – monsoon forest, Aw – savannah. Source: adapted from Beck et al. (2018).

amplitude. For species near the tops of mountains, there may be no place for them to move.

With projected future change, species ranges will expand or contract, the geographic location of ecological zones will shift, forest ecosystem productivity will change, and ecosystems could reorganize following disturbances into ecological systems with no current analogue. Observed shifts in vegetation distribution (Kelly and Goulden, 2008; Lenoir et al., 2010) or increased tree mortality due to drought and heat in forests worldwide (Allen et al., 2010) demonstrate the potential impacts of rapid climate change. These impacts may be aggravated by other human-induced environmental changes such as increases in low elevation ozone concentrations, nitrogenous pollutant deposition, the introduction of exotic insect pests and pathogens, habitat fragmentation, and increased disturbances such as fire (Bernier and Schöne, 2009). Other effects of climate change may also be important for tropical forests. Sea level rise is already impacting on tidal freshwater forests (Doyle et al., 2010) and tidal saltwater forests (mangroves) are expanding landward in subtropical coastal reaches taking over freshwater marsh and forest zones (Di Nitto et al., 2014).

5.1 Forest composition and distribution

Tropical forests are generally highly diverse in species composition. A study in Colombian Amazonia found 1077 tree species with ≥10 cm DBH in 271 genera and 66 families in 95 plots of 0.1 ha (Duivenvoorden, 1996). Studies in Asian tropical rainforests have found comparable levels of tree diversity. Tropical forests are, therefore, inherently more complex, with a vastly higher number of species and more complex interactions between plant and animal species and associated microorganisms, than temperate forests. Ecosystem-process models are used to predict potential future impacts of climate change on tropical forests and calibrating these models requires a sound understanding of individual species and ecosystem responses to change (Mok et al., 2012). Modeling studies use either a statistical or a mechanistic approach. The statistical approach is used to quantify the correlations between climatic variables and species presence/absence or abundance (Guisan and Thuiller, 2005; Araujo and Luoto, 2007). Statistical models are useful in certain situations, but they cannot provide mechanistic explanations of change and are dependent on the time span, spatial extent, and resolution of input data. Mechanistic models may be able to model regeneration responses due to their ability to identify causal changes to climatic and environmental constraints (Kearney and Porter, 2009). However, fully calibrating models for each species in complex tropical forests is unrealistic, and classifying species by traits or life-forms is likely to a more productive approach.

Seedling regeneration may represent a critical process driving climate change impacts. In a study in the Luquillo Forest Dynamics Plot (LFDP), a 16-ha permanent plot in subtropical wet forest in Puerto Rico, Uriarte et al. (2018) found that spatial heterogeneity in soil moisture and conspecific density were the main drivers of seedling survival, with greater survival at low conspecific densities and higher soil moisture. This environmental heterogeneity reduced the impacts of lower rainfall or solar radiation and may buffer or exacerbate impacts of climate fluctuations on forest regeneration. Chidumayo (2005) found that treated tree seedlings in central Zambia performed poorly under warmer climate. On the other hand, in a seasonal moist tropical forest on Barro Colorado Island, Panama, the availability of water increased seedling growth, but did not affect seedling mortality (Bunker and Carson, 2005).

While aspects of climate change may be positive for some tree species in some locations, for example the observed increase in productivity in temperate forests, most projected future changes in climate and their indirect effects are likely to have negative consequences on forest productivity. The sheer variety of leaf traits, stem and root traits, reproductive traits, and water-use traits relevant to water availability in tropical forests means that there is incomplete knowledge about many functional types to enable prediction of the responses of these forests in the face of global climate change (Chaturvedi et al., 2011).

Shifts of species to new localities and loss of species are evident in tropical forests (Zelazowski et al., 2011; D'Antonio, 2000; Campbell et al., 2009; Fischlin et al., 2009). Bernier and Schöne (2009) considered tropical moist forests one of the forest types most vulnerable to climate change. However, there has been much debate about the vulnerability of tropical moist forests (Corlett, 2011; Huntingford et al., 2013; Feeley and Rehm, 2012). Humid tropical forests may be more vulnerable because they are adapted to relatively constant diurnal (between 24 and 27°C) and year-round temperatures (>18°C). Tropical species' physiologies have evolved to function optimally within these narrow temperature ranges, and they may have limited ability to acclimate to global warming compared with temperate ecosystems. Small temperature increases might exceed temperature tolerances for particular species and cause ranges to retract to cooler refuges (Feeley and Silman, 2010; Colwell et al., 2008).

Uncertainties about the response of tropical forests to climate change are associated with differing model representations of vegetation processes and differences in projected temperature and precipitation changes in patterns. Using simulations with 22 climate models and a land surface scheme, Huntingford et al. (2013) found evidence of resilience for tropical forests in Asia, Africa, and the Americas, with only one of the simulations indicating tropical forests would lose biomass by the end of the twenty-first century, and only for the Americas. The largest uncertainties are associated mostly with plant physiological responses and with future emissions scenarios. Uncertainties due to differences in climate projections are significantly smaller. Drier tropical forests may have greater ability to adapt ecologically to drying environments, with studies indicating that these forest types have increased their deciduous species abundance and generally changed more functionally than forests growing in wetter conditions (Aguirre-Gutiérrez et al., 2019).

Large-scale circulation changes are also likely to impact on tropical forests. Using a coupled regional model, Cook and Vizy (2008) found that, under atmospheric CO_2 levels of 757 ppmv by 2081–2100, there would be a 70% reduction in the extent of the Amazon rain forest by the end of the twenty-first century. This was associated with a large eastward expansion of the drier caatinga vegetation that is prominent in the Nordeste region of Brazil today. These changes in vegetation are related to reductions in annual mean rainfall and a modification of the seasonal cycle that are associated with a weakening of tropical circulation systems. In a similar study in Central America, Lyra et al. (2017) found that, even under a low emissions pathway, some areas of tropical rainforest are replaced by savannah and grassland and that under both high- and low-emission scenarios, precipitation will decline and temperatures and dry spells will increase, leading to reduced areas of tropical forests and overall reduced net primary productivity, despite the biomass increases in some areas of Costa Rica and Panama.

A study of the potential impacts of climate change on forests in Costa Rica indicated that climate change is likely to cause shifts in the distribution of tropical forest life zones. High elevation life zones were shown to be more sensitive to changes in temperature, while lower elevation life zones are more sensitive to changes in precipitation. Regional life zone diversity was greatly reduced in an extreme wet and warm climate scenario. Endemic species tend to be clustered in three montane rain forest and the lowland seasonally dry forest life zones and these are particularly sensitive to climate change (Enquist, 2002). But the impacts of climate change on forest distributions will be determined, to a large extent, by the capacity of tree species to move. Many tropical species have complex mutual interactions for pollination and produce large, fleshy fruited seeds that are dispersed primarily by birds or mammals. These interactions limit their capacity to disperse over large distances to new environments.

Fadrique et al. (2019) found that that tropical and subtropical tree communities in the Andes are experiencing directional shifts in composition toward having greater relative abundances of species from lower, warmer elevations, but the rates of compositional change are not uniform across elevations. This variation is probably because of different warming rates and/or the presence of specialized tree communities at ecotones, for example at the transitions between distinct habitats, such as at the timberline or ecotone with cloud forest.

Forests in seasonally dry areas might be more vulnerable due to projected decreases and higher variability in rainfall and potential increases in fire disturbances in these regions. However, tree cover and biomass in tropical and subtropical savannas have been observed to increase over the past century leading to increased carbon storage due to a combination of rising CO_2, climate variability, and climate change (Smith et al., 2014).

In a study of the impacts of climate change on the global distribution of humid tropical forests, based primarily on hydrological determinants, Zelazowski et al. (2011) found some risk of forest retreat, especially in eastern Amazonia, Central America, and parts of Africa, but also a potential for expansion in other regions, for example in the Congo Basin. Potential resilient and vulnerable zones vary at a local scale. CO_2-related reduction in plant water demand lowers the risk of die-back and can lead to possible niche expansion in many regions. Physiological effects of higher temperature may moderate these potential effects.

5.2 Forest functions and ecosystem processes

Changes are expected in ecological relationships and eco-physiological functions of forests (Gonzalez et al., 2010; Fischlin et al., 2009). Phenology of tropical forests, including flowering and fruiting and shedding of leaves, is responding to climate change (Corlett and LaFrankie, 1998; Stork et al., 2007). Cheesman and Winter (2013) studied impacts of elevated daytimes and night-time temperatures on tropical lowland *Ficus insipida* and *Ochroma pyramidale* in controlled-environment chambers at a constant daytime temperature (33°C) and a range of increasing night-time temperatures, complemented by an outdoor open-top chamber study in which night-time temperatures were elevated by 2.4°C above ambient. All species had optimum growth rates under well-watered conditions. However, under the highest temperature regime, shade-tolerant species had substantially reduced growth rates but growth in three lowland pioneers showed only a marginal reduction. All species showed optimal growth at temperatures above those currently found in their native range. Pfeifer et al. (2018) found that canopy structure in the tropics is primarily a consequence of forest adaptation to maximum water deficits experienced within a given region. Climate change and changes in drought regimes are likely to affect forest structure and function.

This is supported by the observed effects of a 25-year drying trend and an intense 4-month dry season on composition of tropical moist forest on Barro Colorado Island in Panama. At least 16 species of shrubs and treelets with affinities for moist microhabitats were headed for local extinction in a 50-ha long-term study plot, but these species may have invaded the forest during a wetter period prior to commencement of observations. A severe drought in 1983 caused unusually high tree mortality and might have led to increases in gap-colonizers because it temporarily opened the forest canopy (Condit et al., 1996).

In a different type of study using analysis of over 2400 long-term measurement plots, Bowman et al. (2014) found that there was a peaked response to temperature in temperate and subtropical eucalypt forests, with maximum growth occurring at a MAT of 11°C and maximum temperature of the warmest month of 25–27°C. Lower temperatures directly constrain growth, while high temperatures primarily reduced growth by reducing water availability. Higher temperatures also appeared to exert a direct negative effect on growth. Overall, the productivity of Australia's temperate eucalypt forests could decline substantially as the climate warms, given that 87% of these forests currently experience a MAT above their 'optimal' temperature.

Esquivel-Muelbert et al. (2019) undertook a widespread study across the Amazon forests of three traits hypothesized to be affected by increasing CO_2 and changing precipitation: maximum tree size, water deficit affiliation of tree genera and wood density. Their results suggest that increased atmospheric CO_2 is driving a shift within tree communities to large-statured species, but long generation times of tropical trees mean that change in composition is relatively slow. Most forest plots have experienced an intensified dry season. There has been no detectable change in mean wood density or water deficit affiliation in the canopy species at the community level. There has been greater observed change among new recruits in plots where the dry season has intensified, with dry-affiliated genera becoming more abundant and increased mortality of genera associated with wetter conditions. They suggest that a slow shift to a dry-affiliated Amazonia is underway.

Large tropical trees may be more vulnerable to increased droughts than smaller ones. Using a large-scale, long-term silvicultural experiment in a transitional Amazonian forest in Bolivia, Shenkin et al. (2018) found tree mortality in both logged and unlogged plots increased in response to drought during the 2004/2005 ENSO event in that region. Hydraulic factors related to tree height were the cause of increased mortality of large trees in unlogged plots. In logged plots, neither tree height nor crown exposure was associated with drought-induced mortality. At a regional scale, Phillips et al. (2008) found that prolonged drought in the Amazon during 2005 contributed to a decline in aboveground biomass and a release of 4.40–5.87 $GtCO_2$.

The greatest threats to tropical forest biodiversity may lie in the synergistic interactions between climate change and human land use. The relative impacts between these drivers are likely to vary by region. Asner et al. (2010) found that in the Amazon, up to 81% of the region may be susceptible to rapid vegetation change, and in the Congo, logging and climate change could negatively affect 35–74% of the basin. Climate-driven changes may play a smaller role in Asia-Oceania, but current land use renders 60–77% of Asia-Oceania susceptible to major biodiversity changes. Aridification could increase access or suitability for cropping of currently nonarable land. These effects could be ameliorated by protected area expansion along key ecological gradients, regulation of human-lit fires, strategic forest–carbon financing and re-evaluation of agricultural and biofuel subsidies (Brodie et al., 2012).

6 Future tropical forest management: mitigation and adaptation to climate change

6.1 Reducing greenhouse gas emissions from tropical forests

As indicated previously, deforestation and degradation of tropical forests are significant contributors to GHG emissions and human-induced climate change. The Paris Agreement on climate change at COP21 was a major landmark in climate policy, providing a comprehensive and inclusive framework for action with ambitious long-term targets to avoid dangerous anthropogenic global warming. Forest-based measures to reduce GHG emissions such as reducing deforestation and increasing sequestration through better management of existing forests and restoring forests were a key element in the Agreement. If fully achieved, these could cut GHG emissions by almost a third (Seymour and Busch, 2016). Over 60 countries referred to REDD+ in their Intended Nationally Determined Contributions and, as part of the Lima Paris Action Agenda, heads of government from major forest countries and partners committed to action prior to the COP meeting to promote equitable rural development, reverse deforestation, and massively increase forest restoration.

Brazil and Norway renewed their US$1 billion partnership to reduce deforestation until 2020 and Germany, Norway, and the UK announced US$5 billion to support country-based REDD+ programs between 2015 and 2020. While the focus is on results-based payments, the prospects for a global market that involves the transfer of emission reduction credits from developing to developed countries are unclear. With tropical developing countries committing to NDCs, they may want to use forest-based emission reductions for their own targets. In the short term, finance for REDD+ or other forest-related emission reduction activities may mainly come through bilateral funds (like the agreements Norway has with Brazil and Indonesia) or multilateral funds such as the World Bank's Forest Carbon Partnership Facility or the Green Climate Fund, which recently committed US$500 million to purchase REDD+ emission reductions. Commitments to reduce emissions from international aviation may create a large new market for forest-based offsets (Hamrick and Gallant, 2018).

Addressing deforestation requires a strategic, integrated approach to agriculture, forestry, and other natural resource and economic development policies. The effects of deforestation can be reduced through the following actions: (1) effective land use planning that involves local people and industries to ensure that deforestation only occurs in areas where conservation and community values will not be unduly affected, (2) enforcement of laws and regulations that prevent forest conversion, (3) education measures to ensure that communities and companies understand the value of intact forests, (4) supporting sustainable forest management efforts that enable communities to reap financial benefits from forests, and (5) providing financial incentives to countries, industries, communities, or households to encourage forest conservation.

Public policies have had a significant impact by reducing deforestation rates in some tropical countries (Smith et al., 2014). However, while the contribution from deforestation is reducing, the contribution from degradation is increasing (Federici et al., 2015). The REDD+ mechanism under the United Nations Framework Convention on Climate Change (UNFCCC) is supporting improved policies and laws, increasing measurement, and monitoring capacity and funds to countries or communities to reduce deforestation. Gumpenberger et al. (2010) concluded that the protection of forests under forest conservation programs (including REDD) could increase carbon uptake in many tropical countries, mainly due to CO_2 fertilization effects, even under changing climate conditions. However, others argue that mitigation benefits from deforestation reduction under REDD+ could be reversed due to increased fire events, and climate-induced feedbacks (Arcidiacono-Bársony et al., 2011).

Reduced impact logging (RIL) can potentially halve carbon emissions by reducing logging degradation in tropical forests (Ellis et al., 2019). However, it is likely to be necessary to move beyond RIL to substantially increase carbon storage by developing more sophisticated, planned forest management schemes with silvicultural treatments that ensure regeneration establishment, post establishment release, and extended rotations of new stands. Research on rainforest silviculture has also focused on more productive forests in higher rainfall areas in the wet and semievergreen tropics, with less in montane or seasonally dry forests where much of the degradation is occurring (Del Cid-Liccardi et al., 2012).

Agricultural companies play an increasingly important role as forest clearance for agriculture shifts from small-scale farmers to agribusinesses. Global demand for agricultural products will continue to rise with increasing human population, but technological improvements can increase food and fiber supplies through increasing productivity without increased deforestation (Richards et al., 2019). Many companies have committed to deforestation-free supply chains to reduce their effects on forests. As mentioned previously, signatories to the New York Declaration on Forests (UN, 2014) committed to halving the rate of loss of natural forests globally by 2020 and striving to end natural forest loss by 2030. This would be achieved through eliminating deforestation from producing key agricultural commodities, reducing deforestation derived from other economic sectors by 2020, and supporting alternatives to deforestation driven by basic needs.

Restrictions to agricultural expansion due to forest conservation, increased energy crop area, afforestation, and reforestation may increase costs of agricultural production and food prices (Smith et al., 2015). Trade liberalization can lead to lower costs of food, but also increases the pressure on land, especially on tropical forests (Schmitz et al., 2012). Trade restrictions and tariffs, such as those currently in place between the United States and China, may also exacerbate forest conversion (Fuchs et al., 2019).

6.2 Tropical forest management for adapting to climate change

Policy on tropical forests and global climate change has so far focused mostly on mitigation, with less emphasis on how management activities may help forest ecosystems adapt to this change (Keenan, 2015). With a certain degree of human-induced climate change locked in to the climate system, and likely future larger-scale changes, adaptation measures will be needed to maintain the productive capacity and resilience of tropical forests. Adaptation options can be anticipatory or reactive, and either planned or autonomous (Prowse and Scott, 2008). They can aim to build resistance to change (e.g. to protect rare, high-value species in a specific location, or a plantation forest that is close to rotation age), or to promote resilience to enable forests to respond to future change while maintaining or providing for the recovery of important ecological processes (Millar et al., 2007).

Climate change assessment and planning needs to shift from simply projecting impacts to evaluating adaptation options. For agriculture and land use in general, there has been a shift in thinking from designing prescriptive adaptation options to identifying management principles as a basis for generating locally specific options. These principles focus on system resilience, such as redundancy, flexibility, and cross-scale awareness. Adaptation options for forests often reflect what is considered currently to be good risk management practice (Innes et al., 2009; Keenan and Nitschke, 2016). This approach is likely to be effective under moderate climate change (Dovers, 2009) but could limit planning for transformational changes (Smith et al., 2011). A longer-term view can provide pathways for adaptation options under more extreme levels of climate change as they emerge in the future (Barnett et al., 2014; Wise et al., 2014).

Adaptation decisions need to consider both the biophysical capacity of forests to respond to climate change and the vulnerability and adaptive capacity of the human societies that depend on forests (Keenan, 2015). Communities and livelihoods are exposed to multiple stressors, and climate risk management strategies need to consider the full set of hazards and risks and compounding socioeconomic stressors in an integrated way. Many consider that other pressures and threats to tropical forests are more pressing and immediate than the long-term implications of climate change. In a survey of tropical forest managers, Guariguata et al. (2012) found that respondents perceived that natural and planted forests are at risk from being affected by climate change. However, they were unsure of the value of investing in adaptation. Climate change ranked below other threats to forests such as commercial agriculture and unplanned logging. Long-term forest planning and management was not a key consideration given other major drivers of forest loss and degradation.

In a study of communities impacted by drought in the forest zone of Cameroon, Bele et al. (2013) identified adaptive strategies such as community-created firebreaks to protect their forests and farms from forest fires, the culture of maize and other vegetables in dried swamps, diversifying income activities or changing food regimes. However, these coping strategies were incommensurate with the rate and magnitude of change being experienced and, therefore, no longer seen as useful. Some adaptive actions, while effective, were resource-inefficient and potentially translated pressure from one sector to another or generated other secondary effects that made them undesirable.

The diversity of conditions in the tropics, the range of potential threats and impacts, and the diversity of human interactions with tropical forests means that adaptation options will need to vary according to location, climate, and the nature of future risks. Adaptation is a process best addressed at local levels, with organizations, communities, businesses, households, or individuals considering their future climate risks and the benefits and costs of different risk management options (Ciurean et al., 2013). This local action will still require coordination across jurisdictions and levels of government, particularly in information provision and management of cross-boundary issues, for example in flood risk, fire management, or species conservation.

Proposed measures to address climate change impacts on biodiversity conservation including addressing preexisting stressors on biodiversity, better preparing for the effects of major natural disturbances, significantly improving off-reserve conservation efforts (including fostering appropriate connectivity), and enhancing the existing reserve system by making it more comprehensive, adequate, and representative (Lindenmayer et al., 2010). For production forest managers, landowners, and producers, adaptation actions in forest management can be grouped into broader land management options, site-specific silvicultural practices, building social and community skills, and policy and planning options (Keenan, 2017).

However, the adaptation options decided on, and the weight adaptation given in decision-making, will in large part depend on cost. There is a risk that some longer-term adaptation options that may limit for instance storm, drought, or pest risks will be substituted by shorter-term or more limited measures (Andersson and Keskitalo, 2018). Adaptation policy or strategy decisions in forest will also be influenced by similar factors to those for other decisions. Adaptation options need to be framed in a way that actors in forestry see as relevant, the focus in the industry, and the nature of available options (Mackay et al., 2017). The nature of institutions, laws, and policies, land tenure, and forest ownership arrangements in different countries will impact on policy design for adaptation and on the willingness to implement options for longer-term adaptation (Keenan et al., 2019).

Intensively managed tree plantations are becoming the dominant source of industrial wood supply (Payn et al., 2015) and plantation area is increasing rapidly in the tropics. These plantations usually consist of a single exotic fast growing species, often using clonal production systems that, while highly productive, might expose the estate to more risks from increased disturbance (wind or fire), pests, or disease. The short rotations (5-10 years) used in these systems provide some capacity for more rapid adaptation, with options to change species or genotypes, or vary management in response to rapidly changing conditions. Increasing diversity in tropical plantations has been proposed as a management strategy for some time to increase resilience, enhance productivity, and provide for a wider range of values (Keenan et al., 1999). Converting monospecific plantations to mixed stands may improve stand stability and reduce increasing abiotic and biotic disturbances due to climate change. However, as indicated earlier, little is known about the extent to which tropical tree species or tropical tree communities can resist increasing disturbances in the short term, such as water limitations due to increasing dry season intensity or length (Kunert and Cárdenas, 2015).

New types of knowledge partnerships can support improved interaction between researchers and policy makers and better decision-making (Preston et al., 2015). The USDA Forest Service has actively promoted these partnerships (Joyce et al., 2009; Nagel et al., 2017), and the experience in the tropical El Yunque National Forest (EYNF) in Puerto Rico suggests that managers are better positioned to address increasing uncertainty and surprise at multiple scales. However, ongoing commitment is required to the resources necessary for implementing adaptive, collaborative forest management, and to provide space and flexibility required to make swift adjustments in response to rapid or unexpected system changes (McGinley, 2017).

Adaptation is a continuous process of 'adapting well' to ongoing change (Tompkins et al., 2010). This requires organizational learning based on experience, new knowledge, and a comprehensive analysis of future options. This can take place through 'learning by doing' or through a process of search and planned modification of routines (Berkhout et al., 2006). However, interpreting climate signals is not easy for organizations, the evidence of change is ambiguous, and the stimuli are not often experienced directly within the organization. For example, many forest managers in Australia felt little need to change practices to adapt to climate change, given both weak policy signals and limited perceived immediate evidence of increasing climate impacts (Cockfield et al., 2011). To explain and predict adaptation to climate change, the combination of personal experience and beliefs must be considered (Blennow, 2012). Adaptation will be as much focused on what we are prepared to lose, and preparing as a society to address that (Barnett et al., 2016) as on what we might want to keep in a future climate.

Adaptation measures may help maintain the mitigation potential of tropical forests. For example, projects that prevent fires and restore degraded forest ecosystems also prevent release of GHGs and enhance carbon stocks. Mitigation and adaptation benefits can also contribute to sustainable development considerations (Louman et al., 2019). Broadly, however, there has been little integration to date of mitigation and adaptation objectives in climate policy. For example, there is little connection between policies supporting REDD+ initiatives and adaptation. Integrating adaptation into REDD+ can advance climate change mitigation goals and objectives for sustainable forest management (Long, 2013). Kant and Wu (2012) considered that adaptation actions in tropical forests (protection against fire and disease, ensuring adequate regeneration and protecting against coastal impacts and desertification) will improve future forest resilience and have significant climate change mitigation value. Discourses on the future of forests in the Congo Basin focus on the opportunities for forests under the REDD+ mechanism, but also recognize the need for the forests and dependent societies and sectors to adapt to potential climate risks. Different agents frame climate change adaptation and mitigation policies in the region in different ways: mitigation only, adaptation only, or an integrated approach. These framings result in differing views on costs and benefits, scale of operation, effectiveness, financial resources, and implementation mechanisms. Overall, the mitigation discourse seems to be stronger than the adaptation discourse (Somorin et al., 2012).

Adaptation is about making good decisions for the future, considering the implications of climatic and other environmental and social change. Tropical forest management will have to adapt to a changing and highly variable climate to effectively provide a role in mitigation, deliver associated ecosystem services, and provide benefits in poverty reduction (Eliasch, 2008; Keenan, 2015). In achieving this, the roles and responsibilities of different levels of government, the private sector, and different parts of the community are still being defined.

Integrated measures might support improved forest and biodiversity conservation, and mixed-species forestry-based afforestation may help maintain or enhance carbon stocks, while also providing adaptation options to enhance resilience of forest ecosystems to climate change (Smith et al., 2015). Integrated approaches between forestry and agriculture might also support both climate mitigation and adaptation policy objectives. Enhancement of soil carbon stocks and integrating trees into agricultural systems have the potential to increase productivity, diversity, and resilience to climate change (Verchot et al., 2007; Smith and Olesen, 2010). It is increasingly recognized that reducing emissions and increasing carbon stocks needs a landscape approach. At the 2018 Global Landscapes Forum, held prior to the UNFCCC COP, participants emphasized the need to engage multiple stakeholders from across different tenures and land uses and consider the multiple benefits and services that people want from the landscapes in developing climate solutions.

7 Future trends

Research on forests and climate change is still strongly focused on assessment of future impacts, responses, and vulnerability of species and ecosystems (and in some cases communities and forest industries) to climate change. There has been some movement from a static view of climate based on long-term averages to a more detailed understanding of the drivers of different climate systems and how these affect the factors of greatest influence on different forest ecosystems processes, such as variability and extremes in temperature or precipitation or fire disturbance. This section provides a brief summary of the research needs related to tropical forests and climate change.

Key knowledge gaps in relation to climate impacts require new research that focuses on individual trees rather than leaf level physiological studies, on long-term rather than short-term changes, and on understanding mechanisms instead of documenting changes. Tree-ring analyses, stable-isotope analyses, manipulative field experiments, and well-validated simulation models can improve predictions of forest responses to global change (Zuidema et al., 2013).

Bonal et al. (2016) identified key gaps to study the potential impacts of drought on tropical forests. This included (i) modelling studies or large-scale manipulation studies combining soil drought, increase in CO_2 concentrations and temperature, or combining soil drought and nutrient levels (ii) relating species- and ecosystem-level responses to extreme droughts; (iii) understanding the role of microbial communities in drought tolerances (iv) considering the impacts of repetitive drought events; (v) assessing the adaptive potential of tropical rainforest tree species to droughts; (vi) improving the ability of ecosystem or earth models to simulate drought responses; and (vii) evaluating the long-term impacts of extreme events. These developments can support simulation of the future of these ecosystems under diverse climate scenarios and predict future carbon balances.

While we are gaining a better understanding of the drivers of tropical forest loss (de Sy et al., 2015; Dezécache et al., 2017) in different settings, more research is required on the policy measures to reduce deforestation, including the appropriate mix of regulation, incentives, and education measures. Bebbington et al. (2018) argue research needs to be conducted to change the terms of public debate about forests and forest communities, and to increase capacities and credibility of institutions that defend the rights of forest users. Similarly, in achieving forest restoration, research on degradation and/or deforestation drivers may inform the visioning phase by identifying opportunities and obstacles (Stanturf et al., 2019).

Many have proposed the landscape approach to restoration (Sayer et al., 2013), but achieving this in practice has been challenging given the complex socio-ecological processes occurring in tropical landscapes and the nature of political organization of national or subnational governance units. These limit capacity to build strong and coherent institutions that span the landscape, leading to fragmented policies that do not match landscape multi-functionality. Research is required to explore landscape governance options to connect administratively fragmented landscapes and to building bridges among actors and sectors. Planning for long-term management of restored (or restoring) landscapes needs attention to create greater resilience in the face of changing climate, resistance to future degradation, and social benefit from provisioning various goods and services (Stanturf et al., 2019). More broadly, a greater focus is required on adaptation to climate change in tropical forests and integrating adaptation and mitigation objectives (Long, 2013; Locatelli et al., 2015). For example, restoring forests in landscapes may increase crop resilience to drought (Wang et al., 2019).

8 Conclusion

Climate change is a major global challenge and tropical forests are integral to addressing this challenge. Tropical forests both contribute to, and can be part of the solution for, mitigating and adapting to climate change. Tropical forests are a major store of carbon and in recent decades, tropical forests have contributed to GHG emissions due to deforestation and forest degradation, but the contribution of tropical forests to the global carbon cycle is currently about neutral, with intact and recovering forests taking up carbon. This is likely to change in the near term due to continued forest loss and effects of climate change on tropical forest. By reducing deforestation in tropical forests, managing forests more sustainably, and increasing carbon stocks through forest restoration, tropical forest landscapes can be a major contributor to global emission reduction objectives.

Tropical forests, and the human communities that depend on them, are diverse, with varying sensitivity and vulnerability to climate change. Given this diversity, and the low science base in many tropical countries, there is still relatively limited understanding of the potential impacts of climate change on tropical forests or on forest-dependent communities. Greater investment in research can support increased understanding, improved management, and increased potential for mitigation and adaptation to climate change.

Tropical forest managers will face multiple challenges in a rapidly changing climate. This analysis indicates that tropical forests and their dependent communities are exposed and vulnerable to climate change, and both forests and people have less capacity to adapt than in other regions. While there are encouraging signs in the global policy arena, and some commitment of funds, the science to support improved decision-making is lacking and there has been relatively little policy development to support adaptation to climate change in tropical forests. Effective forest-based measures to meet the Paris objectives will need to be strongly supported and coordinated with policy frameworks and solutions that mobilize and meet the needs of local actors across multiple-use landscapes.

We are at a critical juncture with tropical forests and climate change. The political attention in the Paris Agreement can provide the platform for large-scale, long-term investment in sustainably managed tropical forests (Seymour and Busch, 2016). Many of the requirements are in place: international consensus on the importance of tropical forests in meeting climate objectives, the framework for REDD+; forest-rich countries have made commitments to protect or restore forests, private companies have committed to reducing tropical deforestation. A broad base of civil society groups, including representatives of indigenous peoples, are supporting better forest conservation and management. The technology is in place to make countries and companies accountable. Increased investment in research and in incentives to protect and better manage tropical forests can make a major contribution to reducing emissions, and limiting the impact of climate change, while supporting capacity of tropical forests, and forest-dependent people, to adapt to future changes in tropical climates.

9 Where to look for further information

Many research centres and universities around the world are involved in research on tropical forests and climate change. Some key organisations include:

The Centre for International Forestry Research (CIFOR https://www.cifor.or g/forests-and-climate-change/) has a comprehensive program on forests and climate change.

The Global Carbon Project (https://www.globalcarbonproject.org/) provides the latest statistics on greenhouse gas emissions from different sources and countries.

Forest Trends (https://www.forest-trends.org/topics/forests/) provides information and analysis on forest carbon markets and finance.

The Intergovernmental Panel on Climate Change provides analysis and synthesis to inform policy and decision making on climate change. Their recent report on Climate Change and Land (https://www.ipcc.ch/srccl/) provides a comprehensive picture of the implications of climate change and the potential for sustainable land management to contribute to climate objectives.

The International Institute for Environment and Development (https://www.iied.org/forests) focuses on improving livelihoods for people living in and near forests and managing the impacts of climate change.

The Nature Conservancy has a comprehensive program on forests (https://www.nature.org/media/aboutus/fact_sheet_climate_and_tropical_forests.pdf).

The New York Declaration on Forests is a consortium of governments, industry and NGOs aiming to end deforestation and restore forests (https://forestdeclaration.org/).

World Agroforestry Centre (ICRAF http://www.worldagroforestry.org/en vironmental-services) has a comprehensive research program on incorporating trees in agriculture to improve productivity, sustainability and resilience.

Members of the International Union of Forest Research Organisations are active around the world in research on forests and climate change. For example, see the taskforce on transforming forest landscapes - https://www.iufro.org/sci ence/task-forces/transforming-forest-landscapes/

The World Resources Institute provides tools for monitoring and assessment of tree and forest change (https://www.wri.org/our-work/topics/forests).

WWF International (https://wwf.panda.org/our_work/forests/) identifies solutions to increase forest protected areas, bring more forests under improved management, halt deforestation and restore degraded forest landscapes.

UN Food and Agriculture Organisation (FAO) provides global assessments of forest resources (http://www.fao.org/forest-resources-asseUNssment/en/) and other analysis and support for forests and climate change.

The UN REDD program (https://www.un-redd.org/) supports nationally led REDD+ processes and promotes stakeholder involvement, including indigenous peoples and other forest-dependent communities, in national and international REDD+ implementation.

The World Bank Group (https://www.worldbank.org/en/topic/forests#1) supports countries to harness the potential of forests to reduce poverty, better integrate forests into their economies, and protect and strengthen the environmental role of forests. The Forest Carbon Partnership Facility (https://www.forestcarbonpartnership.org/) supports country readiness for REDD+ and pilots results-based payments to countries that have advanced through REDD+ readiness and implementation and have achieved verifiable emission reductions in their forest and broader land-use sectors.

10 References

Aguirre-Gutiérrez, J., Oliveras, I., Rifai, S., Fauset, S., Adu-Bredu, S., Affum-Baffoe, K., Baker, T. R., Feldpausch, T. R., Gvozdevaite, A., Hubau, W., Kraft, N. J. B., Lewis, S. L., Moore, S., Niinemets, Ü, Peprah, T., Phillips, O. L., Ziemińska, K., Enquist, B. and Malhi, Y. 2019. Drier tropical forests are susceptible to functional changes in response to a long-term drought. *Ecology Letters* 22(5), 855–65. doi:10.1111/ele.13243.

Alkama, R. and Cescatti, A. 2016. Biophysical climate impacts of recent changes in global forest cover. *Science* 351(6273), 600–4. doi:10.1126/science.aac8083.

Allen, C. D., Macalady, A. K., Chenchouni, H., Bachelet, D., McDowell, N., Vennetier, M., Kitzberger, T., Rigling, A., Breshears, D. D., Hogg, E. H., Gonzalez, P., Fensham, R., Zhang, Z., Castro, J., Demidova, N., Lim, J., Allard, G., Running, S. W., Semerci, A. and Cobb, N. 2010. A global overview of drought and heat-induced tree mortality reveals emerging climate change risks for forests. *Forest Ecology and Management* 259(4), 660–84. doi:10.1016/j.foreco.2009.09.001.

Andersson, E. and Keskitalo, E. C. H. 2018. Adaptation to climate change? Why business-as-usual remains the logical choice in Swedish forestry. *Global Environmental Change* 48, 76–85. doi:10.1016/j.gloenvcha.2017.11.004.

Arcidiacono-Bársony, C., Ciais, P., Viovy, N. and Vuichard, N. 2011. REDD mitigation. *Procedia Environmental Sciences* 6, 50–9. doi:10.1016/j.proenv.2011.05.006.

Asner, G. P., Loarie, S. R. and Heyder, U. 2010. Combined effects of climate and land-use change on the future of humid tropical forests. *Conservation Letters* 3(6), 395–403. doi:10.1111/j.1755-263X.2010.00133.x.

Baccini, A., Goetz, S. J., Walker, W. S., Laporte, N. T., Sun, M., Sulla-Menashe, D., Hackler, J., Beck, P. S. A., Dubayah, R., Friedl, M. A., Samanta, S. and Houghton, R. A. 2012. Estimated carbon dioxide emissions from tropical deforestation improved by carbon-density maps. *Nature Climate Change* 2(3), 182–5. doi:10.1038/nclimate1354.

Barnett, J., Graham, S., Mortreux, C., Fincher, R., Waters, E. and Hurlimann, A. 2014. A local coastal adaptation pathway. *Nature Climate Change* 4(12), 1103–8. doi:10.1038/nclimate2383.

Barnett, J., Tschakert, P., Head, L. and Adger, W. N. 2016. A science of loss. *Nature Climate Change* 6(11), 976–8. doi:10.1038/nclimate3140.

Bebbington, A. J., Bebbington, D. H., Sauls, L. A., Rogan, J., Agrawal, S., Gamboa, C., Imhof, A., Johnson, K., Rosa, H., Royo, A., Toumbourou, T. and Verdum, R. 2018. Resource extraction and infrastructure threaten forest cover and community rights. *Proceedings of the National Academy of Sciences of the United States of America* 115(52), 13164–73. doi:10.1073/pnas.1812505115.

Beck, H. E., Zimmermann, N. E., McVicar, T. R., Vergopolan, N., Berg, A. and Wood, E. F. 2018. Present and future Köppen-Geiger climate classification maps at 1-km resolution. *Scientific Data* 5, 180214.

Bele, M. Y., Tiani, A. M., Somorin, O. A. and Sonwa, D. J. 2013. Exploring vulnerability and adaptation to climate change of communities in the forest zone of Cameroon. *Climatic Change* 119(3–4), 875–89. doi:10.1007/s10584-013-0738-z.

Berkhout, F., Hertin, J. and Gann, D. M. 2006. Learning to adapt: organisational adaptation to climate change impacts. *Climatic Change* 78(1), 135–56. doi:10.1007/s10584-006-9089-3.

Bernier, P. and Schöne, D. 2009. Adapting forests and their management to climate change: an overview. *Unasylva* 60, 5–11.

Blennow, K. 2012. Adaptation of forest management to climate change among private individual forest owners in Sweden. *Forest Policy and Economics* 24, 41–7. doi:10.1016/j.forpol.2011.04.005.

Bonal, D., Burban, B., Stahl, C., Wagner, F. and Hérault, B. 2016. The response of tropical rainforests to drought–lessons from recent research and future prospects. *Annals of Forest Science* 73, 27–44. doi:10.1007/s13595-015-0522-5.

Bonan, G. B. 2008. Forests and climate change: forcings, feedbacks, and the climate benefits of forests. *Science* 320(5882), 1444–9. doi:10.1126/science.1155121.

Bowman, D. M. J. S., Williamson, G. J., Keenan, R. J. and Prior, L. D. 2014. A warmer world will reduce tree growth in evergreen broadleaf forests: evidence from Australian temperate and subtropical eucalypt forests. *Global Ecology and Biogeography* 23(8), 925–34. doi:10.1111/geb.12171.

Brodie, J., Post, E. and Laurance, W. F. 2012. Climate change and tropical biodiversity: a new focus. *Trends in Ecology and Evolution* 27(3), 145–50. doi:10.1016/j.tree.2011.09.008.

Bunker, D. E. and Carson, W. P. 2005. Drought stress and tropical forest woody seedlings: effect on community structure and composition. *Journal of Ecology* 93(4), 794–806. doi:10.1111/j.1365-2745.2005.01019.x.

Campbell, E. M., Saunders, S. C., Coates, K. D., Meidinger, D. V., MacKinnon, A., O'Neill, G. A., MacKillop, D. J., DeLong, S. C. and Morgan, D. G. 2009. *Ecological Resilience and Complexity: A Theoretical Framework for Understanding and Managing British Columbia's Forest Ecosystems in a Changing Climate*. British Columbia, Ministry of Forests and Range, Forest Science Program, Victoria, BC.

Chaturvedi, R. K., Gopalakrishnan, R., Jayaraman, M., Bala, G., Joshi, N. V., Sukumar, R. and Ravindranath, N. H. 2011. Impact of climate change on Indian forests: a dynamic vegetation modeling approach. *Mitigation and Adaptation Strategies for Global Change* 16(2), 119–42. doi:10.1007/s11027-010-9257-7.

Cheesman, A. W. and Winter, K. 2013. Elevated night-time temperatures increase growth in seedlings of two tropical pioneer tree species. *The New Phytologist* 197(4), 1185–92. doi:10.1111/nph.12098.

Chidumayo, E. N. 2005. Effects of climate on the growth of exotic and indigenous trees in central Zambia. *Journal of Biogeography* 32(1), 111–20. doi:10.1111/j.1365-2699.2004.01130.x.

Ciurean, R. L., Schröter, D. and Glade, T. 2013. Conceptual Frameworks of Vulnerability Assessments for Natural Disasters Reduction. In: Tiefenbacher, J. P. (Ed.), *Approaches to Disaster Management - Examining the Implications of Hazards, Emergencies and Disasters*. IntechOpen.

Clark, D. A. 2007. Detecting tropical forests' responses to global climatic and atmospheric change: current challenges and a way forward. *Biotropica* 39(1), 4–19. doi:10.1111/j.1744-7429.2006.00227.x.

Cockfield, G., Maraseni, T., Buys, L., Sommerfeld, J., Wilson, C. and Athukorala, W. 2011. *Socioeconomic Implications of Climate Change with Regard to Forests and Forest Management. Contribution of Work Package 3 to the Forest Vulnerability Assessment*. National Climate Change Adaptation Research Facility, Gold Coast, Australia, p. 105.

Colwell, R. K., Brehm, G., Cardelús, C. L., Gilman, A. C. and Longino, J. T. 2008. Global warming, elevational range shifts, and lowland biotic attrition in the wet tropics. *Science* 322(5899), 258–61. doi:10.1126/science.1162547.

Condit, R., Hubbell, S. P. and Foster, R. B. 1996. Changes in tree species abundance in a Neotropical forest: impact of climate change. *Journal of Tropical Ecology* 12(2), 231–56. doi:10.1017/S0266467400009433.

Cook, K. H. and Vizy, E. K. 2008. Effects of twenty-first-century climate change on the Amazon rain forest. *Journal of Climate* 21(3), 542–60. doi:10.1175/2007JCLI1838.1.

Corlett, R. T. 2011. Impacts of warming on tropical lowland rainforests. *Trends in Ecology and Evolution (Personal Edition)* 26, 606–13. doi:10.1016/j.tree.2011.06.015.

Corlett, R. T. 2012. Climate change in the tropics: the end of the world as we know it? *Biological Conservation* 151(1), 22–5. doi:10.1016/j.biocon.2011.11.027.

Corlett, R. T. and LaFrankie, J. V. 1998. Potential impacts of climate change on tropical Asian forests through an influence on phenology. *Climatic Change* 39(2/3), 439–53. doi:10.1023/A:1005328124567.

Cusack, D. F., Karpman, J., Ashdown, D., Cao, Q., Ciochina, M., Halterman, S., Lydon, S. and Neupane, A. 2016. Global change effects on humid tropical forests: evidence for biogeochemical and biodiversity shifts at an ecosystem scale. *Reviews of Geophysics* 54(3), 523–610. doi:10.1002/2015RG000510.

D'Antonio, C. M. 2000. Fire, plant invasions, and global changes. In: Mooney, H. A. and Hobbs, R. J. (Eds), *Invasive Species in a Changing World*. Island Press, Washington DC, pp. 65–93.

Del Cid-Liccardi, C., Kramer, T., Ashton, M. S. and Griscom, B. 2012. Managing carbon sequestration in tropical forests. In: Ashton, M., Tyrrell, M., Spalding, D. and Gentry B. (Eds), *Managing Forest Carbon in a Changing Climate*. Springer, Dordrecht, pp. 183–204.

de Sy, V., Herold, M., Achard, F., Beuchle, R., Clevers, J. G. P. W., Lindquist, E. and Verchot, L. 2015. Land use patterns and related carbon losses following deforestation in South America. *Environmental Research Letters* 10(12).

Dezécache, C., Salles, J. M., Vieilledent, G. and Hérault, B. 2017. Moving forward socio-economically focused models of deforestation. *Global Change Biology* 23(9), 3484–500. doi:10.1111/gcb.13611.

Di Nitto, D., Neukermans, G., Koedam, N., Defever, H., Pattyn, F., Kairo, J. G. and Dahdouh-Guebas, F. 2014. Mangroves facing climate change: landward migration potential in response to projected scenarios of sea level rise. *Biogeosciences* 11(3), 857–71. doi:10.5194/bg-11-857-2014.

Dovers, S. 2009. Normalizing adaptation. *Global Environmental Change* 19(1), 4–6. doi:10.1016/j.gloenvcha.2008.06.006.

Doyle, T. W., Krauss, K. W., Conner, W. H. and From, A. S. 2010. Predicting the retreat and migration of tidal forests along the northern Gulf of Mexico under sea-level rise. *Forest Ecology and Management* 259(4), 770–7. doi:10.1016/j.foreco.2009.10.023.

Dressler, W., Wilson, D., Clendenning, J., Cramb, R., Mahanty, S., Lasco, R., Keenan, R., To, P. and Gevana, D. 2015. Examining how long fallow swidden systems impact upon livelihood and ecosystem services outcomes compared with alternative land-uses in the uplands of Southeast Asia. *Journal of Development Effectiveness* 7(2), 210–29. doi:10.1080/19439342.2014.991799.

Duivenvoorden, J. F. 1996. Patterns of tree species richness in rain forests of the middle Caqueta Area, Colombia, NW Amazonia. *Biotropica* 28(2), 142–58. doi:10.2307/2389070.

Ellis, P. W., Gopalakrishna, T., Goodman, R. C., Putz, F. E., Roopsind, A., Umunay, P. M., Zalman, J., Ellis, E. A., Mo, K., Gregoire, T. G. and Griscom, B. W. 2019. Reduced-impact logging for climate change mitigation (RIL-C) can halve selective logging emissions from tropical forests. *Forest Ecology and Management* 438, 255–66. doi:10.1016/j.foreco.2019.02.004.

Enquist, C. A. F. 2002. Predicted regional impacts of climate change on the geographical distribution and diversity of tropical forests in Costa Rica. *Journal of Biogeography* 29(4), 519–34. doi:10.1046/j.1365-2699.2002.00695.x.

Erb, K.-H., Kastner, T., Luyssaert, S., Houghton, R. A., Kuemmerle, T., Olofsson, P. and Haberl, H. 2013. Bias in the attribution of forest carbon sinks. *Nature Climate Change* 3(10), 854–6. doi:10.1038/nclimate2004.

Esquivel-Muelbert, A., Baker, T. R., Dexter, K. G., Lewis, S. L., Brienen, R. J. W., Feldpausch, T. R., Lloyd, J., Monteagudo-Mendoza, A., Arroyo, L., Álvarez-Dávila, E., Higuchi, N., Marimon, B. S., Marimon-Junior, B. H., Silveira, M., Vilanova, E., Gloor, E., Malhi, Y., Chave, J., Barlow, J., Bonal, D., Davila Cardozo, N., Erwin, T., Fauset, S., Hérault, B., Laurance, S., Poorter, L., Qie, L., Stahl, C., Sullivan, M. J. P., Ter Steege, H., Vos, V. A., Zuidema, P. A., Almeida, E., Almeida de Oliveira, E., Andrade, A., Vieira, S. A., Aragão, L., Araujo-Murakami, A., Arets, E., Aymard C, G. A., Baraloto, C., Camargo, P. B., Barroso, J. G., Bongers, F., Boot, R., Camargo, J. L., Castro, W., Chama Moscoso, V., Comiskey, J., Cornejo Valverde, F., Lola da Costa, A. C., Del Aguila Pasquel, J., Di Fiore, A., Fernanda Duque, L., Elias, F., Engel, J., Flores Llampazo, G., Galbraith, D., Herrera Fernández, R., Honorio Coronado, E., Hubau, W., Jimenez-Rojas, E., Lima, A. J. N., Umetsu, R. K., Laurance, W., Lopez-Gonzalez, G., Lovejoy, T., Aurelio Melo Cruz, O., Morandi, P. S., Neill, D., Núñez Vargas, P., Pallqui Camacho, N. C., Parada

Gutierrez, A., Pardo, G., Peacock, J., Peña-Claros, M., Peñuela-Mora, M. C., Petronelli, P., Pickavance, G. C., Pitman, N., Prieto, A., Quesada, C., Ramírez-Angulo, H., Réjou-Méchain, M., Restrepo Correa, Z., Roopsind, A., Rudas, A., Salomão, R., Silva, N., Silva Espejo, J., Singh, J., Stropp, J., Terborgh, J., Thomas, R., Toledo, M., Torres-Lezama, A., Valenzuela Gamarra, L., van de Meer, P. J., van der Heijden, G., van der Hout, P., Vasquez Martinez, R., Vela, C., Vieira, I. C. G. and Phillips, O. L. 2019. Compositional response of Amazon forests to climate change. *Global Change Biology* 25(1), 39–56. doi:10.1111/gcb.14413.

Federici, S., Tubiello, F. N., Salvatore, M., Jacobs, H. and Schmidhuber, J. 2015. New estimates of CO2 forest emissions and removals: 1990–2015. *Forest Ecology and Management* 352, 89–98. doi:10.1016/j.foreco.2015.04.022.

Feeley, K. J. and Silman, M. R. 2010. Land-use and climate change effects on population size and extinction risk of Andean plants. *Global Change Biology* 16(12), 3215–22. doi:10.1111/j.1365-2486.2010.02197.x.

Feeley, K. J. and Rehm, E. M. 2012. Amazon's vulnerability to climate change heightened by deforestation and man-made dispersal barriers. *Global Change Biology* 18(12), 3606–14. doi:10.1111/gcb.12012.

Fischlin, A., Ayres, M., Karnosky, D., Kellomäki, S., Louman, B., Ong, C., Plattner, G. K., Santoso, H., Thompson, I., Booth, T. H., Marcar, N., Scholes, B., Swanston, C. and Zamolodchikov, D. 2009. Future environmental impacts and vulnerabilities. In: Seppälä, R., Buck, A. and Katila, P. (Eds), *Adaptation of Forests and People to Climate Change: A Global Assessment Report*. IUFRO World Series, Helsinki, pp. 53–100.

Fuchs, R., Alexander, P., Brown, C., Cossar, F., Henry, R. C. and Rounsevell, M. 2019. *Why The US-China Trade War Spells Disaster for the Amazon*. Nature Publishing Group.

Gonzalez, P., Neilson, R. P., Lenihan, J. M. and Drapek, R. J. 2010. Global patterns in the vulnerability of ecosystems to vegetation shifts due to climate change. *Global Ecology and Biogeography* 19(6), 755–68. doi:10.1111/j.1466-8238.2010.00558.x.

Guariguata, M. R., Locatelli, B. and Haupt, F. 2012. Adapting tropical production forests to global climate change: risk perceptions and actions. *International Forestry Review* 14(1), 27–38. doi:10.1505/146554812799973226.

Gumpenberger, M., Vohland, K., Heyder, U., Poulter, B., Macey, K., Rammig, A., Popp, A. and Cramer, W. 2010. Predicting pan-tropical climate change induced forest stock gains and losses—implications for REDD. *Environmental Research Letters* 5(1), 014013. doi:10.1088/1748-9326/5/1/014013.

Hamrick, M. and Gallant, M. 2018. *Fertile Ground: State of Forest Carbon Finance*. Forest Trends, Washington DC, p. 88.

Huang, M. and Asner, G. P. 2010. Long-term carbon loss and recovery following selective logging in Amazon forests. *Global Biogeochemical Cycles* 24(3). doi:10.1029/2009GB003727.

Huntingford, C., Zelazowski, P., Galbraith, D., Mercado, L. M., Sitch, S., Fisher, R., Lomas, M., Walker, A. P., Jones, C. D., Booth, B. B. B., Malhi, Y., Hemming, D., Kay, G., Good, P., Lewis, S. L., Phillips, O. L., Atkin, O. K., Lloyd, J., Gloor, E., Zaragoza-Castells, J., Meir, P., Betts, R., Harris, P. P., Nobre, C., Marengo, J. and Cox, P. M. 2013. Simulated resilience of tropical rainforests to CO2-induced climate change. *Nature Geoscience* 6(4), 268–73. doi:10.1038/ngeo1741.

Innes, J., Joyce, L. A., Kellomäki, S., Louman, B., Ogden, A., Thompson, I., Ayres, M., Ong, C., Santoso, H., Sohngen, B. and Wreford, A. 2009. Management for adaptation. In:

Seppälä, R., Buck, A. and Katila, P. (Eds), *Adaptation of Forests and People to Climate Change: A Global Assessment Report*. IUFRO, Helsinki, pp. 135–86.

IPCC. 2013. *Climate Change 2013: the Physical Science Basis. Contribution of Working Group I to the Fifth Assessment Report of the Intergovernmental Panel on Climate Change*. Stocker, T. F., Qin, D., Plattner, G.-K., Tignor, M., Allen, S. K., Boschung, J., Nauels, A., Xia, Y., Bex, V. and Midgley, P. M. (Eds). IPCC, Cambridge, UK and New York, NY, p. 1535.

Jolly, W. M., Cochrane, M. A., Freeborn, P. H., Holden, Z. A., Brown, T. J., Williamson, G. J. and Bowman, D. M. 2015. Climate-induced variations in global wildfire danger from 1979 to 2013. *Nature Communications* 6, 7537. doi:10.1038/ncomms8537.

Joyce, L. A., Blate, G. M., McNulty, S. G., Millar, C. I., Moser, S., Neilson, R. P. and Peterson, D. L. 2009. Managing for multiple resources under climate change: national forests. *Environmental Management* 44(6), 1022–32. doi:10.1007/s00267-009-9324-6.

Kant, P. and Wu, S. 2012. Should adaptation to climate change be given priority over mitigation in tropical forests? *Carbon Management* 3(3), 303–11. doi:10.4155/cmt.12.29.

Keenan, R. J. 2015. Climate change impacts and adaptation in forest management: a review. *Annals of Forest Science* 72, 145–167.

Keenan, R. J. 2017. Climate change and Australian production forests: impacts and adaptation. *Australian Forestry* 80(4), 197–207. doi:10.1080/00049158.2017.1360170.

Keenan, R. J. and Nitschke, C. 2016. Forest management options for adaptation to climate change: a case study of tall, wet eucalypt forests in Victoria's Central Highlands region. *Australian Forestry* 79(2), 96–107. doi:10.1080/00049158.2015.1130095.

Keenan, R. J., Lamb, D., Parrotta, J. and Kikkawa, J. 1999. Ecosystem management in tropical timber plantations. *Journal of Sustainable Forestry* 9(1–2), 117–34. doi:10.1300/J091v09n01_10.

Keenan, R. J., Reams, G. A., Achard, F., de Freitas, J. V., Grainger, A. and Lindquist, E. 2015. Dynamics of global forest area: results from the FAO Global Forest Resources Assessment 2015. *Forest Ecology and Management* 352, 9–20. doi:10.1016/j.foreco.2015.06.014.

Keenan, R. J., Nelson, H., Keskitalo, E. C. H. and Bergh, J. 2019. Climate change adaptation in forest production systems in a globalizing economy. In: Preston, B. L. and Keskitalo, E. C. H. (Eds), *Research Handbook on Climate Change Adaptation Policy*. Edward Elgar Publishing, Cheltenham.

Kelly, A. E. and Goulden, M. L. 2008. Rapid shifts in plant distribution with recent climate change. *Proceedings of the National Academy of Sciences of the United States of America* 105(33), 11823–6. doi:10.1073/pnas.0802891105.

Kunert, N. and Cárdenas, A. M. 2015. Are mixed tropical tree plantations more resistant to drought than monocultures? *Forests* 6(12), 2029–46. doi:10.3390/f6062029.

Le Quéré, C., Moriarty, R., Andrew, R. M., Peters, G. P., Ciais, P., Friedlingstein, P., Jones, S. D., Sitch, S., Tans, P., Arneth, A., Boden, T. A., Bopp, L., Bozec, Y., Canadell, J. G., Chini, L. P., Chevallier, F., Cosca, C. E., Harris, I., Hoppema, M., Houghton, R. A., House, J. I., Jain, A. K., Johannessen, T., Kato, E., Keeling, R. F., Kitidis, V., Klein Goldewijk, K., Koven, C., Landa, C. S., Landschützer, P., Lenton, A., Lima, I. D., Marland, G., Mathis, J. T., Metzl, N., Nojiri, Y., Olsen, A., Ono, T., Peng, S., Peters, W., Pfeil, B., Poulter, B., Raupach, M. R., Regnier, P., Rödenbeck, C., Saito, S., Salisbury, J. E., Schuster, U.,

Schwinger, J., Séférian, R., Segschneider, J., Steinhoff, T., Stocker, B. D., Sutton, A. J., Takahashi, T., Tilbrook, B., van der Werf, G. R., Viovy, N., Wang, Y.-P., Wanninkhof, R., Wiltshire, A. and Zeng, N. 2015. Global carbon budget 2014. *Earth System Science Data* 7(1), 47–85. doi:10.5194/essd-7-47-2015.

Lenoir, J., Gegout, J. C., Dupouey, J. L., Bert, D. and Svenning, J.-C. 2010. Forest plant community changes during 1989–2007 in response to climate warming in the Jura Mountains (France and Switzerland). *Journal of Vegetation Science* 21(5), 949–64. doi:10.1111/j.1654-1103.2010.01201.x.

Lindenmayer, D. B., Steffen, W., Burbidge, A. A., Hughes, L., Kitching, R. L., Musgrave, W., Stafford Smith, M. and Werner, P. A. 2010. Conservation strategies in response to rapid climate change: Australia as a case study. *Biological Conservation* 143(7), 1587–93. doi:10.1016/j.biocon.2010.04.014.

Liu, Y. Y., van Dijk, A. I. J. M., de Jeu, R. A. M., Canadell, J. G., McCabe, M. F., Evans, J. P. and Wang, G. 2015. Recent reversal in loss of global terrestrial biomass. *Nature Climate Change* 5(5), 470–4. doi:10.1038/nclimate2581.

Liu, J., Bowman, K. W., Schimel, D. S., Parazoo, N. C., Jiang, Z., Lee, M., Bloom, A. A., Wunch, D., Frankenberg, C., Sun, Y., O'Dell, C. W., Gurney, K. R., Menemenlis, D., Gierach, M., Crisp, D. and Eldering, A. 2017. Contrasting carbon cycle responses of the tropical continents to the 2015-2016 El Nino. *Science* 358(6360). doi:10.1126/science.aam5690.

Long, A. 2013. REDD plus, Adaptation, and sustainable forest management: toward effective polycentric global forest governance. *Tropical Conservation Science* 6(3), 384–408. doi:10.1177/194008291300600306.

Louman, B., Keenan, R. J., Kleinschmit, D., Atmadja, S., Sitoe, A. A., Nhantumbo, I., de Camino Velozo, R. and Morales, J. P. 2019. SDG 13: Climate Action–Impacts on Forests and People, in: Katila, P., Pierce-Colfer, C., de Jong, W., Galloway, G., Pacheco, P., Winkel, G. (Eds.), Sustainable Development Goals: Their Impacts on Forests and People. *Cambridge University Press, Cambridge* 419–44.

Lyra, A., Imbach, P., Rodriguez, D., Chou, S. C., Georgiou, S. and Garofolo, L. 2017. Projections of climate change impacts on Central America tropical rainforest. *Climatic Change* 141(1), 93–105. doi:10.1007/s10584-016-1790-2.

Mackay, H., Keskitalo, E. C. H. and Pettersson, M. J. B. I. 2017. Getting invasive species on the political agenda: agenda setting and policy formulation in the case of ash dieback in the UK. *Biological Invasions* 19(7), 1953–70. doi:10.1007/s10530-017-1415-3.

McGinley, K. 2017. Adapting tropical forest policy and practice in the context of the anthropocene: opportunities and challenges for the El Yunque national forest in Puerto Rico. *Forests* 8(7). doi:10.3390/f8070259.

Meyfroidt, P., Roy Chowdhury, R., de Bremond, A., Ellis, E. C., Erb, K.-H., Filatova, T., Garrett, R. D., Grove, J. M., Heinimann, A., Kuemmerle, T., Kull, C. A., Lambin, E. F., Landon, Y., le Polain de Waroux, Y., Messerli, P., Müller, D., Nielsen, J. Ø., Peterson, G. D., Rodriguez García, V., Schlüter, M., Turner, B. L. and Verburg, P. H. 2018. Middle-range theories of land system change. *Global Environmental Change* 53, 52–67. doi:10.1016/j.gloenvcha.2018.08.006.

Millar, C. I., Stephenson, N. L. and Stephens, S. L. 2007. Climate change and forests of the future: managing in the face of uncertainty. *Ecological Applications: a Publication of the Ecological Society of America* 17(8), 2145–51. doi:10.1890/06-1715.1.

Mitchard, E. T. A. 2018. The tropical forest carbon cycle and climate change. *Nature* 559(7715), 527–34. doi:10.1038/s41586-018-0300-2.

Mok, H.-F., Arndt, S. K. and Nitschke, C. R. 2012. Modelling the potential impact of climate variability and change on species regeneration potential in the temperate forests of South-Eastern Australia. *Global Change Biology* 18(3), 1053–72. doi:10.1111/j.1365-2486.2011.02591.x.

Montagnini, F. and Jordan, C. F. 2005. Classification of tropical forests. In: Montagnini, F. and Jordan, C. F. *Tropical Forest Ecology. Tropical Forestry.* Springer, Berlin, Heidelberg 75–96.

Nagel, L. M., Palik, B. J., Battaglia, M. A., D'Amato, A. W., Guldin, J. M., Swanston, C. W., Janowiak, M. K., Powers, M. P., Joyce, L. A., Millar, C. I., Peterson, D. L., Ganio, L. M., Kirschbaum, C. and Roske, M. R. 2017. Adaptive silviculture for climate change: a national experiment in manager-scientist partnerships to apply an adaptation framework. *Journal of Forestry* 115(3), 167–78. doi:10.5849/jof.16-039.

Pan, Y., Birdsey, R. A., Fang, J., Houghton, R., Kauppi, P. E., Kurz, W. A., Phillips, O. L., Shvidenko, A., Lewis, S. L., Canadell, J. G., Ciais, P., Jackson, R. B., Pacala, S. W., McGuire, A. D., Piao, S., Rautiainen, A., Sitch, S. and Hayes, D. 2011. A large and persistent carbon sink in the world's forests. *Science* 333(6045), 988–93. doi:10.1126/science.1201609.

Payn, T., Carnus, J.-M., Freer-Smith, P., Kimberley, M., Kollert, W., Liu, S., Orazio, C., Rodriguez, L., Silva, L. N. and Wingfield, M. J. 2015. Changes in planted forests and future global implications. *Forest Ecology and Management* 352, 57–67. doi:10.1016/j.foreco.2015.06.021.

Pfeifer, M., Gonsamo, A., Woodgate, W., Cayuela, L., Marshall, A. R., Ledo, A., Paine, T. C. E., Marchant, R., Burt, A., Calders, K., Courtney-Mustaphi, C., Cuni-Sanchez, A., Deere, N. J., Denu, D., de Tanago, J. G., Hayward, R., Lau, A., Macía, M. J., Olivier, P. I., Pellikka, P., Seki, H., Shirima, D., Trevithick, R., Wedeux, B., Wheeler, C., Munishi, P. K. T., Martin, T., Mustari, A. and Platts, P. J. 2018. Tropical forest canopies and their relationships with climate and disturbance: results from a global dataset of consistent field-based measurements. *Forest Ecosystems* 5(1). doi:10.1186/s40663-017-0118-7.

Phillips, O. L., Lewis, S. L., Baker, T. R., Chao, K. J. and Higuchi, N. 2008. The changing Amazon forest. *Philosophical Transactions of the Royal Society of London. Series B, Biological Sciences* 363(1498), 1819–27. doi:10.1098/rstb.2007.0033.

Preston, B. L., Suppiah, R., Macadam, I., et al. 2006. Climate change in the Asia/Pacific region: a consultancy report prepared for the Climate Change and Development Roundtable. CSIRO Marine and Atmospheric Research, Aspendale, p. 93.

Preston, B. L., Rickards, L., Fünfgeld, H. and Keenan, R. J. 2015. Toward reflexive climate adaptation research. *Current Opinion in Environmental Sustainability* 14, 127–35. doi:10.1016/j.cosust.2015.05.002.

Prowse, M. and Scott, L. 2008. Assets and adaptation: an emerging debate. *IDS Bulletin* 39(4), 42–52. doi:10.1111/j.1759-5436.2008.tb00475.x.

Richards, M., Arslan, A., Cavatassi, R. and Rosenstock, T. S. 2019. Climate change mitigation potential of agricultural practices supported by IFAD investments. International Fund for Agriculture Development, Rome.

Sayer, J., Sunderland, T., Ghazoul, J., Pfund, J. L., Sheil, D., Meijaard, E., Venter, M., Boedhihartono, A. K., Day, M., Garcia, C., van Oosten, C. and Buck, L. E. 2013. Ten principles for a landscape approach to reconciling agriculture, conservation, and other competing land uses. *Proceedings of the National Academy of Sciences of the United States of America* 110(21), 8349-56. doi:10.1073/pnas.1210595110.

Schmitz, C., Biewald, A., Lotze-Campen, H., Popp, A., Dietrich, J. P., Bodirsky, B., Krause, M. and Weindl, I. 2012. Trading more food: implications for land use, greenhouse gas emissions, and the food system. *Global Environmental Change* 22(1), 189-209. doi:10.1016/j.gloenvcha.2011.09.013.

Seymour, F. and Busch, J. 2016. *Why Forests? Why Now? The Science, Economics, and Politics of Tropical Forests and Climate Change*. Center for Global Development, Washington DC.

Shenkin, A., Bolker, B., Peña-Claros, M., Licona, J. C., Ascarrunz, N. and Putz, F. E. 2018. Interactive effects of tree size, crown exposure and logging on drought-induced mortality. *Philosophical Transactions of the Royal Society of London. Series B, Biological Sciences* 373(1760). doi:10.1098/rstb.2018.0189.

Smith, P. and Olesen, J. E. 2010. Synergies between the mitigation of, and adaptation to, climate change in agriculture. *The Journal of Agricultural Science* 148(5), 543-52. doi:10.1017/S0021859610000341.

Smith, P., Bustamante, M., Ahammad, H., Clark, H., Dong, H., Elsiddig, E. A., Haberl, H., Harper, R., House, J. and Jafari, M. 2014. Agriculture, forestry and other land use (AFOLU). *Climate Change 2014: Mitigation of Climate Change. Contribution of Working Group III to the Fifth Assessment Report of the Intergovernmental Panel on Climate Change*. Cambridge University Press.

Somorin, O. A., Brown, H. C. P., Visseren-Hamakers, I. J., Sonwa, D. J., Arts, B. and Nkem, J. 2012. The Congo Basin forests in a changing climate: policy discourses on adaptation and mitigation (REDD+). *Global Environmental Change* 22(1), 288-98. doi:10.1016/j.gloenvcha.2011.08.001.

Stanturf, J. A., Kleine, M., Mansourian, S., Parrotta, J., Madsen, P., Kant, P., Burns, J. and Bolte, A. 2019. Implementing forest landscape restoration under the Bonn Challenge: a systematic approach. *Annals of Forest Science* 76(2), 50. doi:10.1007/s13595-019-0833-z.

Stork, N. E., Balston, J., Farquhar, G. D., Franks, P. J., Holtum, J. A. M. and Liddell, M. J. 2007. Tropical rainforest canopies and climate change. *Austral Ecology* 32(1), 105-12. doi:10.1111/j.1442-9993.2007.01741.x.

Tompkins, E. L., Adger, W. N., Boyd, E., Nicholson-Cole, S., Weatherhead, K. and Arnell, N. 2010. Observed adaptation to climate change: UK evidence of transition to a well-adapting society. *Global Environmental Change* 20(4), 627-35. doi:10.1016/j.gloenvcha.2010.05.001.

UN. 2014. Forests action statements and action plan. Available at: http://www.un.org/climatechange/summit/wp-content/uploads/sites/2/2014/07/New-York-Declaration-on-Forest-%E2%80%93-Action-Statement-and-Action-Plan.pdf.

Uriarte, M., Muscarella, R. and Zimmerman, J. K. 2018. Environmental heterogeneity and biotic interactions mediate climate impacts on tropical forest regeneration. *Global Change Biology* 24(2), e692-704. doi:10.1111/gcb.14000.

Verchot, L. V., Noordwijk, M., Kandji, S., Tomich, T., Ong, C., Albrecht, A., Mackensen, J., Bantilan, C., Anupama, K. V. and Palm, C. 2007. Climate change: linking adaptation and mitigation through agroforestry. *Mitigation and Adaptation Strategies for Global Change* 12(5), 901-18. doi:10.1007/s11027-007-9105-6.

Wang, Y., Huang, J. and Chen, X. 2019. Do forests relieve crop thirst in the face of drought? *Empirical Evidence from South China* 55, 105-14.

Wise, R. M., Fazey, I., Stafford Smith, M., Park, S. E., Eakin, H. C., Archer Van Garderen, E. R. M. and Campbell, B. 2014. Reconceptualising adaptation to climate change as part of pathways of change and response. *Global Environmental Change* 28, 325-36. doi:10.1016/j.gloenvcha.2013.12.002.

Zelazowski, P., Malhi, Y., Huntingford, C., Sitch, S. and Fisher, J. B. 2011. Changes in the potential distribution of humid tropical forests on a warmer planet. *Philosophical Transactions. Series A, Mathematical, Physical, and Engineering Sciences* 369(1934), 137-60. doi:10.1098/rsta.2010.0238.

Zuidema, P. A., Baker, P. J., Groenendijk, P., Schippers, P., van der Sleen, P., Vlam, M. and Sterck, F. 2013. Tropical forests and global change: filling knowledge gaps. *Trends in Plant Science* 18(8), 413-9. doi:10.1016/j.tplants.2013.05.006.

www.ingramcontent.com/pod-product-compliance
Lightning Source LLC
Chambersburg PA
CBHW050715280326
41926CB00088B/3037

'What are the core personal attitudes and professional skills necessary to work as a Jungian analyst? The editors have invited some of the most senior practitioners and influential thinkers in the Jungian canon to tease out what each considers to be irreducible, essential psychotherapeutic abilities. The results of this international reflective process are richly rewarding and multi-faceted, identifying key elements of the analytic attitude in different societies and playing with the art of communication in creative tension between con-sciousness and the unknown. Each chapter adds a fresh insight and all dem-onstrate the deep satisfaction felt in helping patients and trainees alike to explore the symbolic inner world of the unconscious mind.'

Catherine Crowther, *Training Analyst,*
Society of Analytical Psychology, London

'By putting at the center the specific skills a Jungian Analyst must acquire and develop to fulfill the tasks and objectives of Jungian analysis, this book is an indispensable reference for educational activities in all training programs. It is an excellent contribution to the understanding of the specificity of Jungian practice.'

Pilar Amezaga, *Clinical Professor of*
Jungian Psychotherapy, Catholic University of
Uruguay and Vice President of the IAAP

'As analytical psychology has moved from a small movement largely within Western Europe and North America to a global approach to analysis and psy-chotherapy, as well as facing a decisive generational transition as the last direct connections to the founding generation of analysts fade, defining the founda-tional characteristics of Jung's approach to the psyche becomes increasingly important for our understanding of analytical psychology's unique contri-bution to mental and spiritual health in the contemporary world. The core competencies discussed in this volume directly address this growing need and represent an indispensable resource for training the next generation of ana-lytical psychologists.'

George B. Hogenson, **PhD**, *Chicago Society of*
Analytical Psychologists

'A book exploring the core competencies of the psychotherapist is badly needed. A row of senior analysts give a deep understanding of the effective-ness of Jungian psychotherapy. As the role of the psychotherapist has been undeservedly underestimated this book is highly recommendable for anybody who enables another person's growth.'

Kathrin Asper, PhD, *Training Analyst and*
Supervisor at ISAP